Lecture Notes in Computer Science 3940

Commenced Publication in 1973
Founding and Former Series Editors:
Gerhard Goos, Juris Hartmanis, and Jan van Leeuwen

Craig Saunders   Marko Grobelnik
Steve Gunn   John Shawe-Taylor (Eds.)

Subspace, Latent Structure and Feature Selection

Statistical and Optimization
Perspectives Workshop, SLSFS 2005
Bohinj, Slovenia, February 23-25, 2005
Revised Selected Papers

 Springer

Volume Editors

Craig Saunders
Steve Gunn
John Shawe-Taylor
University of Southampton
School of Electronics and Computer Science
ISIS Group
Southampton, SO17 1BJ, UK
E-mail: {cjs,srg,jst}@ecs.soton.ac.uk

Marko Grobelnik
J. Stefan Institute
Department of Intelligent Systems
Jamova 39, 1000 Ljubljana, Slovenia
E-mail: Marko.Grobelnik@ijs.si

Library of Congress Control Number: 2006925463

CR Subject Classification (1998): F.1, F.2, I.2.6-8, I.2.3, I.4-6, F.4.1, H.3

LNCS Sublibrary: SL 1 – Theoretical Computer Science and General Issues

ISSN   0302-9743
ISBN-10   3-540-34137-4 Springer Berlin Heidelberg New York
ISBN-13   978-3-540-34137-6 Springer Berlin Heidelberg New York

Springer is a part of Springer Science+Business Media

springer.com

© Springer-Verlag Berlin Heidelberg 2006
Printed in Germany

Typesetting: Camera-ready by author, data conversion by Scientific Publishing Services, Chennai, India
Printed on acid-free paper   SPIN: 11752790   06/3142   5 4 3 2 1 0

Preface

The inspiration for this volume was a workshop held under the auspices of the PASCAL Network of Excellence. Details of the event and more information about the Network can be found under the heading 'Workshop Organization.' The aim of this preface is to provide an overview of the contributions to this volume, placing this research in its wider context.

The aim of the workshop was to bring together researchers working on subspace and latent variable techniques in different research communities in order to create bridges and enable cross-fertilization of ideas. For this reason we deliberately sought invited refereed contributions that would survey a broader field of research giving a common notation and entry point into the individual papers.

The five invited contributions are (in alphabetical order of first author) Avrim Blum on *Random Projection, Margins, Kernels and Feature Selection*, Wray Buntine and Aleks Jakulin on *Discrete Principal Components Analysis*, Dunja Mladenić on *Dimensionality Reduction by Feature Selection in Machine Learning*, Roman Rosipal and Nicole Krämer on *Overview and Recent Advances in Partial Least Squares*, and Mike Titterington on *Some Aspects of Latent Structure Analysis*.

Blum considers subspace selection by random projection. The theoretical analysis of this approach provides an important bound on the generalization of large margin algorithms, but it can also be implemented in kernel defined feature spaces through a two-stage process. The paper provides a survey of a number of clean and important theoretical results. Buntine and Jakulin consider method of determining latent structure based on probabilistic generative models of the data. Their paper gives an introduction to these advanced and effective methods presented from within the machine learning community. Titterington's contribution is a closely related paper but comes from the statistics tradition providing a general framework within which discrete and continuous combinations of latent and observed variables can be placed. Mladenić considers the restricted class of axis parallel subspaces that correspond to feature selection. There is a long tradition of this approach within machine learning and the paper provides an overview of a range of techniques for selecting features, discussing their weaknesses and carefully evaluating their performance. Rosipal and Krämer give a detailed introduction to partial least squares, an important method of subspace selection developed within the chemometrics research community. It can be thought of as an adaptation of principal components analysis where the projection directions have been chosen to be well-suited for solving a set of regression tasks. The authors discuss the kernelization of the technique together with other more recent results.

The contributed papers cover a range of application areas and technical approaches. Agakov and Barber develop a probabilistic modelling technique with a novel twist of using encoding models rather than generative ones; Monay et al. again consider computer vision using a probabilistic modelling approach; Navot et al. analyze a simple two Gaussian example to show that feature selection can make significant differences in performance and that techniques such as support vector machines are not able to avoid the difficulties of non-informative features in this case; Bouveyron et al. consider a computer vision application using a probabilistic modelling approach; Gruber and Weiss continue the computer vision theme but introduce prior knowledge to enhance the ability to factorize image data to perform 3D reconstruction; Savu-Krohn and Auer use a clustering approach to reduce feature dimensions for image data; Rogers and Gunn consider random forests as an approach to feature selection; Maurer gives frequentist style generalization bounds on PCA-like subspace methods; and finally Reunanen discusses the biases of using cross-validation to do feature selection and outlines some techniques to prevent the introduction of such a bias.

We commend the volume to you as a broad introduction to many of the key approaches that have been developed for subspace identification and feature selection. At the same time the contributed talks give insightful examples of applications of the techniques and highlight recent developments in this rapidly expanding research area. We hope that the volume will help bridge the gaps between different disciplines and hence enable creative collaborations that will bring benefit to all involved.

February 2006

Marko Grobelnik
Steve Gunn
Craig Saunders
John Shawe-Taylor

Workshop Organization

Many of the papers in this proceedings volume were presented at the PASCAL Workshop entitled *Subspace, Latent Structure and Feature Selection Techniques: Statistical and Optimization Perspectives* which took place in Bohinj, Slovenia during February, 23–25 2005.

The workshop was part of a Thematic Programme Linking Learning and Statistics with Optimization that ran over the first half of 2005. The PASCAL Network is a European Network of Excellence funded by the European Union under the IST programme. It currently has around 300 researchers at 55 institutions. Its center of gravity is machine learning, but it aims to build links with both optimization and statistics as well as with a range of application areas. It sponsors and co-sponsors a wide range of workshops either organized independently or co-located with international conferences. More information can be found on the website http://www.pascal-network.org.

The Bohinj workshop was hosted by the Institute Josef Stefan, which provided all of the local organization. We are indebted to them for all of the hard work that they put into making the event such a success, although even they could not have planned the magical winter scene that awaited us on our arrival. Particular thanks are due to Tina Anžič, who handled the reservations and hotel bookings as well as many of the travel arrangements.

This work was supported by the IST Programme of the European Community, under the PASCAL Network of Excellence, IST-2002-506778.

Organizing Committee

Marko Grobelnik	Jožef Stefan Institute, Ljubljana, Slovenia
Steve Gunn	ISIS Group, University of Southampton, UK
Craig Saunders	ISIS Group, University of Southampton, UK
John Shawe-Taylor	ISIS Group, University of Southampton, UK

Table of Contents

Invited Contributions

Contributed Papers

Discrete Component Analysis

Wray Buntine[1] and Aleks Jakulin[2]

[1] Helsinki Institute for Information Technology (HIIT),
Dept. of Computer Science, PL 68,
00014, University of Helsinki, Finland
`Wray.Buntine@hiit.fi`
[2] Department of Knowledge Technologies,
Jozef Stefan Institute, Jamova 39, 1000 Ljubljana, Slovenia
`jakulin@acm.org`

Abstract. This article presents a unified theory for analysis of components in discrete data, and compares the methods with techniques such as independent component analysis, non-negative matrix factorisation and latent Dirichlet allocation. The main families of algorithms discussed are a variational approximation, Gibbs sampling, and Rao-Blackwellised Gibbs sampling. Applications are presented for voting records from the United States Senate for 2003, and for the Reuters-21578 newswire collection.

1 Introduction

Principal component analysis (PCA) [MKB79] is a key method in the statistical engineering toolbox. It is well over a century old, and is used in many different ways. PCA is also known as the Karhünen-Loève transform or Hotelling transform in image analysis, and a variation is latent semantic analysis (LSA) in text analysis [DDL+90]. It is a kind of eigen-analysis since it manipulates the eigen-spectrum of the data matrix. It is usually applied to measurements and real valued data, and used for feature extraction or data summarization. LSA might not perform the centering step (subtracting the mean from each data vector prior to eigen-analysis) on the word counts for a document to preserve matrix sparseness, or might convert the word counts to real-valued `tf*idf` [BYRN99]. The general approach here is *data reduction*.

Independent component analysis (ICA, see [HKO01]) is in some ways an extension of this general approach, however it also involves the estimation of so-called latent, unobservable variables. This kind of estimation follows the major statistical methodology that deals with general unsupervised methods such as clustering and factor analysis. The general approach is called *latent structure analysis* [Tit], which is more recent, perhaps half a century old. The data is modelled in a way that admits unobservable variables, that influence the observable variables. Statistical inference is used to "reconstruct" the unobservable variables from the data jointly with general characteristics of the unobservable variables themselves. This is a theory with particular assumptions (i.e., a "model"), so the method may arrive at poor results.

C. Saunders et al. (Eds.): SLSFS 2005, LNCS 3940, pp. 1–33, 2006.

Relatively recently the statistical computing and machine learning community has become aware of seemingly similar approaches for discrete observed data that appears under many names. The best known of these in this community are probabilistic latent semantic indexing (PLSI) [Hof99], non-negative matrix factorisation (NMF) [LS99] and latent Dirichlet allocation (LDA) [BNJ03]. Other variations are discussed later in Section 5. We refer to these methods jointly as *Discrete Component Analysis* (DCA), and this article provides a unifying model for them.

All the above approaches assume that the data is formed from individual observations (documents, individuals, images), where each observation is described through a number of variables (words, genes, pixels). All these approaches attempt to summarize or explain the similarities between observations and the correlations between variables by inferring latent variables for each observation, and associating latent variables with observed variables.

These methods are applied in the social sciences, demographics and medical informatics, genotype inference, text and image analysis, and information retrieval. By far the largest body of applied work in this area (using citation indexes) is in genotype inference due to the Structure program [PSD00]. A growing body of work is in text classification and topic modelling (see [GS04, BPT04]), and language modelling in information retrieval (see [AGvR03, BJ04, Can04]). As a guide, argued in the next section, the methods apply when PCA or ICA might be used, but the data is discrete.

Here we present in Section 3 a unified theory for analysis of components in discrete data, and compare the methods with related techniques in Section 5. The main families of algorithms discussed in Section 7 are a variational approximation, Gibbs sampling, and Rao-Blackwellised Gibbs sampling. Applications are presented in Section 8 for voting records from the United States Senate for 2003, and the use of components in subsequent classification.

2   Views of DCA

One interpretation of the DCA methods is that they are a way of approximating large sparse discrete matrices. Suppose we have a $500,000$ documents made up of $1,500,000$ different words. A document such as a page out of Dr. Seuss's *The Cat in The Hat*, is first given as a *sequence of words*.

So, as fast as I could, I went after my net. And I said, "With my net I can bet them I bet, I bet, with my net, I can get those Things yet!"

It can be put in the *bag of words* representation, where word order is lost. This yields a list of words and their counts in brackets:

after(1) and(1) as(2) bet(3) can(2) could(1) fast(1) get(1) I(7) my(3) net(3) said(1) so(1) them(1) things(1) those(1) went(1) with(2) yet(1) .

Although the word 'you' never appears in the original, we do not include 'you (0)' in the representation since zeros are suppressed. This sparse vector can be represented as a vector in full word space with $1,499,981$ zeroes and the counts above making the non-zero entries in the appropriate places. Given a matrix made up of rows of such vectors of non-negative integers dominated by zeros, it is called here a *large sparse discrete matrix*.

Bag of words is a basic representation in information retrieval [BYRN99]. The alternative is a sequence of words. In DCA, either representation can be used and the models act the same, up to any word order effects introduced by incremental algorithms. This detail is made precise in subsequent sections.

In this section, we argue from various perspectives that large sparse discrete data is not well suited to standard PCA or ICA methods.

2.1 Issues with PCA

PCA has been normally applied to numerical data, where individual instances are vectors of real numbers. However, many practical datasets are based on vectors of integers, non-negative counts or binary values. For example, a particular word cannot have a negative number of appearances in a document. The vote of a senator can only take three values: Yea, Nay or Not Voting. We can transform all these variables into real numbers using tf*idf, but this is a linear weighting that does not affect the shape of a distribution.

With respect to modelling count data in linguistic applications, Dunning makes the following warning [Dun94]:

> Statistics based on the assumption of normal distribution are invalid in most cases of statistical text analysis unless either enormous corpora are used, or the analysis is restricted to only the very most common words (that is, the ones least likely to be of interest). This fact is typically ignored in much of the work in this field. Using such invalid methods may seriously overestimate the significance of relatively rare events. Parametric statistical analysis based on the binomial or multinomial distribution extends the applicability of statistical methods to much smaller texts than models using normal distributions and shows good promise in early applications of the method.

While PCA is not always considered a method based on Gaussians, it can be justified using Gaussian distributions [Row98, TB99]. Moreover, PCA is justified using a least squares distance measure, and most of the properties of Gaussians follow from the distance measure alone. Rare events correspond to points far away under an L_2 norm.

Fundamentally, there are two different kinds of large sample approximating distributions that dominate discrete statistics: the Poisson and the Gaussian. For instance, a large sample binomial is approximated as a Poisson[1] when the probability is small and as a Gaussian otherwise [Ros89]. Figure 2.1 illustrates this by showing the Gaussian and Poisson approximations to a binomial with

---

[1] This is a distribution on integers where a rate is given for events to occur, and the distribution is over the total number of events counted.

Poisson(100p) and Gaussian(100p,100p(1-p)) approximations to Binomial(100,p)

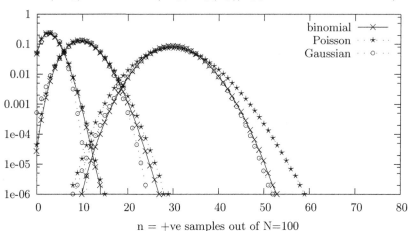

n = +ve samples out of N=100

sample size $N = 100$ for different proportions ($p = 0.03, 0.01, 0.03$). Plots are done with probability in log scale so the errors for low probability values are highlighted. One can clearly see the problem here: the Gaussian provides a reasonable approximate for medium values of the proportion p but for small values it severely underestimates low probabilities. When these low probability events occur, as they always will, the model becomes distorted.

Thus in image analysis based on analogue to digital converters, where data is counts, Gaussian errors can sometimes be assumed, but the Poisson should be used if counts are small. DCA then avoids Gaussian modelling of the data, using a Poisson or multinomial directly.

Another critique of the general style of PCA comes from the psychology literature, this time it is used as a justification for DCA [GS02]. Griffiths and Steyvers argue against the least squares distance of PCA:

> While the methods behind LSA were novel in scale and subject, the suggestion that similarity relates to distance in psychological space has a long history (Shepard, 1957). Critics have argued that human similarity judgments do not satisfy the properties of Euclidean distances, such as symmetry or the triangle inequality. Tversky and Hutchinson (1986) pointed out that Euclidean geometry places strong constraints on the number of points to which a particular point can be the nearest neighbor, and that many sets of stimuli violate these constraints.

They also considered power law arguments which PCA violates for associated words.

2.2   Component Analysis as Approximation

In the data reduction approach for PCA, one seeks to reduce each J-dimensional data vector to a smaller K-dimensional vector. This can be done by approximating the full data matrix as a product of smaller matrices, one representing the

$$\underbrace{\left\{\begin{pmatrix} w_{1,1} & w_{1,2} & \cdots & w_{1,J} \\ w_{2,1} & w_{2,2} & \cdots & w_{2,J} \\ \vdots & \vdots & \ddots & \vdots \\ w_{I,1} & w_{I,2} & \cdots & w_{I,J} \end{pmatrix}\right.}_{\text{data matrix}} \simeq \underbrace{\begin{pmatrix} l_{1,1} & \cdots & l_{1,K} \\ l_{2,1} & \cdots & l_{2,K} \\ \vdots & \ddots & \vdots \\ l_{I,1} & \cdots & l_{I,K} \end{pmatrix}}_{\text{score matrix}} * \underbrace{\left.\begin{pmatrix} \theta_{1,1} & \theta_{2,1} & \cdots & \theta_{J,1} \\ \vdots & \vdots & \ddots & \vdots \\ \theta_{1,K} & \theta_{2,K} & \cdots & \theta_{J,K} \end{pmatrix}\right\}}_{\text{loading matrix}^T}$$

Fig. 1. The matrix approximation view

reduced vectors called the component/factor *score matrix*, and one representing a data independent part called the component/factor *loading matrix*, as shown in Figure 1. In PCA according to least squares theory, this approximation is made by eliminating the lower-order eigenvectors, the least contributing components [MKB79].

If there are I documents, J words and K components, then the matrix on the left has $I * J$ entries and the two matrices on the right have $(I + J) * K$ entries. This represents a simplification when $K \ll I, J$. We can view DCA methods as seeking the same goal in the case where the matrices are sparse and discrete.

When applying PCA to large sparse discrete matrices, or LSA using word count data interpretation of the components, if it is desired, becomes difficult (it was not a goal of the original method [DDL$^+$90]). Negative values appear in the component matrices, so they cannot be interpreted as "typical documents" in any usual sense. This applies to many other kinds of sparse discrete data: low intensity images (such as astronomical images) and verb-noun data used in language models introduced by [PTL93], for instance.

The cost function being minimized then plays an important role. DCA places constraints on the approximating score matrix and loading matrix in Figure 1 so that they are also non-negative. It also uses an entropy distance instead of a least squares distance.

2.3 Independent Components

Independent component analysis (ICA) was also developed as an alternative to PCA. Hyvänen and Oja [HO00] argue that PCA methods merely find uncorrelated components. ICA then was developed as a way of representing multivariate data with truly *independent* components. In theory, PCA approximates this also if the data is Gaussian [TB99], but in practice it rarely is.

The basic formulation is that a K-dimensional data vector $\boldsymbol{w}$ is a linear invertible function of K independent components represented as a K-dimensional latent vector $\boldsymbol{l}$, $\boldsymbol{w} = \boldsymbol{\Theta l}$ for a square invertible matrix $\boldsymbol{\Theta}$. Note the ICA assumes $J = K$ in our notation. $\boldsymbol{\Theta}$ plays the same role as the loading matrix above. For some univariate density model U, the independent components are distributed as

$p(\boldsymbol{l} \,|\, U) \;=\; \prod_k p(l_k \,|\, U)$, thus one can get a likelihood formula $p(\boldsymbol{w} \,|\, \boldsymbol{\Theta}, U)$ using the above equality[2].

The Fast ICA algorithm [HO00] can be interpreted as a maximum likelihood approach based on this model and likelihood formula. In the sparse discrete case, however, this formulation breaks down for the simple reason that $\boldsymbol{w}$ is mostly zeros: the equation can only hold if $\boldsymbol{l}$ and $\boldsymbol{\Theta}$ are discrete as well and thus the gradient-based algorithms for ICA cannot be justified. To get around this in practice, when applying ICA to documents [BKG03], word counts are sometimes first turned into `tf*idf` scores [BYRN99].

To arrive at a formulation more suited to discrete data, we can relax the equality in ICA (i.e., $\boldsymbol{w} = \boldsymbol{\Theta}\boldsymbol{l}$) to be an expectation:

$$\mathbb{E}_{\boldsymbol{w} \sim p(\boldsymbol{w}|\boldsymbol{l},U)} \left[\boldsymbol{w}\right] \;=\; \boldsymbol{\Theta}\boldsymbol{l} \;.$$

We still have independent components, but a more robust relationship between the data and the score vector. Correspondence between ICA and DCA has been noted in [BJ04, Can04]. With this expectation relationship, the dimension of $\boldsymbol{l}$ can now be less than the dimension of $\boldsymbol{w}$, $K < J$, and thus $\boldsymbol{\Theta}$ would be a rectangular matrix.

3   The Basic Model

A good introduction to these models from a number of viewpoints is by [BNJ03, Can04, BJ04]. Here we present a general model. The notation of words, bags and documents will be used throughout, even though other kinds of data representations also apply. In statistical terminology, a word is an observed variable, and a document is a data vector (a list of observed variables) representing an instance. In machine learning terminology, a word is a feature, a bag is a data vector, and a document is an instance. Notice that the bag collects the words in the document and loses their ordering. The bag is represented as a data vector $\boldsymbol{w}$. It is now J-dimensional. The latent, hidden or unobserved vector $\boldsymbol{l}$ called the component *scores* is K-dimensional. The term *component* is used here instead of topic, factor or cluster. The parameter matrix is the previously mentioned component loading matrix $\boldsymbol{\Theta}$, and is $J \times K$.

At this point, it is also convenient to introduce the symbology used throughout the paper. The symbols summarised in Table 1 will be introduced as we go.

3.1   Bags or Sequences of Words?

For a document $\boldsymbol{x}$ represented as a sequence of words, if $\boldsymbol{w} = \text{bag}(\boldsymbol{x})$ is its bagged form, the bag of words, represented as a vector of counts. In the simplest

---

[2] By a change of coordinates

$$p(\boldsymbol{w} \,|\, \boldsymbol{\Theta}, U) \;=\; \frac{1}{\det(\boldsymbol{\Theta})} \prod_k p\left((\boldsymbol{\Theta}^{-1}\boldsymbol{w})_k \,|\, U\right)$$

Table 1. Summary of major symbols

I	number of documents
(i)	subscript to indicate document, sometimes dropped
J	number of different words, size of the dictionary
K	number of components
$L_{(i)}$	number of words in document i
S	number of words in the collection, $\sum_i L_{(i)}$
$\boldsymbol{w}_{(i)}$	vector of J word counts in document i, row totals of $\boldsymbol{V}$, entries $w_{j,(i)}$
$\boldsymbol{c}_{(i)}$	vector of K component counts for document i, column totals of $\boldsymbol{V}$
$\boldsymbol{V}$	matrix of word counts per component, dimension $J \times K$, entries $v_{j,k}$
$\boldsymbol{l}_{(i)}$	vector of K component scores for document i, entries $l_{k,(i)}$
$\boldsymbol{m}_{(i)}$	$\boldsymbol{l}_{(i)}$ normalised, entries $m_{k,(i)}$
$\boldsymbol{k}_{(i)}$	vector of $L_{(i)}$ sequential component assignments for the words in document i, entries $k_{l,(i)} \in [1, \ldots, K]$
$\boldsymbol{\Theta}$	component loading matrix, dimension $J \times K$, entries $\theta_{j,k}$
$\boldsymbol{\theta}_{\cdot,k}$	component loading vector for component k, a column of $\boldsymbol{\Theta}$
$\boldsymbol{\alpha}, \boldsymbol{\beta}$	K-dimensional parameter vectors for component priors

case, one can use a multinomial with sample size $L = |\boldsymbol{x}|$ and vocabulary size $J = |\boldsymbol{w}|$ to model the bag, or alternatively L independent discrete distributions[3] with J outcomes to model each x_l. The bag $\boldsymbol{w}$ corresponds to the sequence $\boldsymbol{x}$ with the order lost, thus there are $\frac{(\sum_j w_j)!}{\prod_j w_j!}$ different sequences that map to the same bag $\boldsymbol{w}$. The likelihoods for these two simple models thus differ by just this combinatoric term.

Note that some likelihood based methods such as maximum likelihood, some Bayesian methods, and some other fitting methods (for instance, a cross validation technique) use the likelihood as a black-box function. They take values or derivatives but otherwise do not further interact with the likelihood. The combinatoric term mapping bag to sequence representations can be ignored here safely because it does not affect the fitting of the parameters for $\mathcal{M}$. Thus for these methods, it is irrelevant whether the data is treated as a bag or as a sequence. This is a general property of multinomial data.

Thus, while we consider bag of words in this article, most of the theory applies equally to the sequence of words representation[4]. Implementation can easily address both cases with little change to the algorithms, just to the data handling routines.

---

[3] The *discrete* distribution is the multivariate form of a Bernoulli where an index $j \in \{0, 1, ..., J - 1\}$ is sampled according to a J-dimensional probability vector.

[4] Some advanced fitting methods such as Gibbs sampling do not treat the likelihood as a black-box. They introduce latent variables that expands the functional form of the likelihood, and they may update parts of a document in turn. For these, ordering effects can be incurred by bagging a document, since updates for different parts of the data will now be done in a different order. But the combinatoric term mapping bag to sequence representations will still be ignored and the algorithms are effectively the same up to the ordering affects.

3.2   General DCA

The general formulation introduced in Section 2.3 is an unsupervised version of a linear model, and it applies to the bag of words $\boldsymbol{w}$ as

$$\mathbb{E}_{\boldsymbol{w}\sim p(\boldsymbol{w}|l,\boldsymbol{\Theta})}\left[\boldsymbol{w}\right] \;=\; \boldsymbol{\Theta}\boldsymbol{l} \tag{1}$$

The expected value (or mean) of the data is given by the dot product of the component loading matrix $\boldsymbol{\Theta}$ and some latent component scores $\boldsymbol{l}$.

In full probability (or Bayesian) modelling [GCSR95], we are required to give a distribution for all the non-deterministic values in a model, including model parameters and the latent variables. In likelihood modelling [CB90], we are required to give a distribution for all the data values in a model, including observed and latent variables. These are the core methodologies in computational statistics, and most others extend these two. The distribution for the data is called a likelihood in both methodologies.

The likelihood of a document is the primary way of evaluating a probabilistic model. Although likelihood is not strictly a probability in classical statistics, we can interpret them as a probability that a probabilistic model $\mathcal{M}$ would generate a document $\boldsymbol{x}$, $P(\boldsymbol{x}|\mathcal{M})$. On the other hand, it is also a way of determining whether the document is usual or unusual: documents with low likelihood are often considered to be outliers or anomalies. If we trust our documents, low likelihoods indicate problems with the model. If we trust out model, a low likelihood indicates problems with a document.

Thus to complete the above formulation for DCA, we need to give distributions matching the constraint in Equation (1), to specify the likelihood. Distributions are needed for:

- how the sequence $\boldsymbol{x}$ or bag $\boldsymbol{w}$ is distributed given its mean $\boldsymbol{\Theta}\boldsymbol{l}$ formed from the component loading matrix,
- how the component scores $\boldsymbol{l}$ are distributed,
- and if full probability modelling is used, how the component loading matrix $\boldsymbol{\Theta}$ is distributed apriori, as well as any parameters.

The formulation of Equation (1) is also called an *admixture model* in the statistical literature [PSD00]. This is in contrast with a *mixture model* [GCSR95] which uses a related constraint

$$\mathbb{E}_{\boldsymbol{w}\sim p(\boldsymbol{w}|l)}\left[\boldsymbol{w}\right] \;=\; \boldsymbol{\theta}_{\cdot,k} \;,$$

for some latent variable k representing the single latent component for $\boldsymbol{w}$. Since k is unobserved, this also corresponds to making a weighted sum of the probability distributions for each $\boldsymbol{\theta}_{\cdot,k}$.

4   The Model Families

This section introduces some forms of DCA using specific distributions for the sequence $\boldsymbol{x}$ or bag $\boldsymbol{w}$ and the component scores $\boldsymbol{l}$. The fundamental model here is the Gamma-Poisson Model (GP model for short). Other models can be presented as variations. The probability for a document is given for each model, both for the case where the latent variables are known (and thus are on the right-hand side), and for the case where the latent variables are included in the left-hand side.

4.1   The Gamma-Poisson Model

The general Gamma-Poisson form of DCA, introduced as GaP [Can04] is now considered in more detail:

- Document data is supplied in the form of *word counts*. The word count for each word type is w_j. Let L be the total count, so $L = \sum_j w_j$.
- The document also has *component scores l* that indicate the amount of the component in the document. These are latent or unobserved. The entries l_k are independent and gamma distributed

$$l_k \sim \text{Gamma}(\alpha_k, \beta_k) \qquad \text{for } k = 1, \ldots, K.$$

The β_k affects scaling of the components[5], while α_k changes the shape of the distribution, shown in Figure 2.

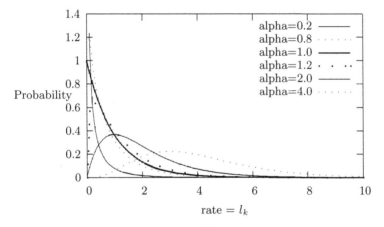

Fig. 2. Gamma distribution for different values of α_k

- There is a *component loading matrix Θ* of size $J \times K$ with entries $\theta_{j,k}$ that controls the partition of features amongst each component. In the matrix, each column for component k is normalised across the features, meaning that $\sum_j \theta_{j,k} = 1$. Thus each column represents the proportions of words/features in component k.
- The observed data w is now Poisson distributed, for each j

$$w_j \sim \text{Poisson}((\Theta l)_j) .$$

---

[5] Conventions for the gamma vary. Sometimes a parameter $1/\beta_k$ is used. Our convention is revealed in Equation (2).

- The K parameters (α_k, β_k) to the gamma distributions give K-dimensional parameter vectors $\boldsymbol{\alpha}$ and $\boldsymbol{\beta}$. Initially these vectors will be treated as constants, and their estimation is the subject of later work.
- When using Bayesian of full probability modelling, a prior is needed for $\boldsymbol{\Theta}$. A Dirichlet prior can be used for each k-th component of $\boldsymbol{\Theta}$ with J prior parameters γ_j, so $\boldsymbol{\theta}_{.,k} \sim \mathrm{Dirichlet}_J(\boldsymbol{\gamma})$. In practice we use a Jeffreys' prior, which has $\gamma_j = 0.5$. The use of a Dirichlet has no strong justification other than being conjugate [GCSR95], but the Jeffreys' prior has some minimax properties [CB94] that make it more robust.

The hidden or latent variables here are the component scores $\boldsymbol{l}$. The model parameters are the gamma parameters $\boldsymbol{\beta}$ and $\boldsymbol{\alpha}$, and the component loading matrix $\boldsymbol{\Theta}$. Denote the model as GP standing for Gamma-Poisson. The full likelihood for each document, $p(\boldsymbol{w}, \boldsymbol{l} \mid \boldsymbol{\beta}, \boldsymbol{\alpha}, \boldsymbol{\Theta}, K, \mathrm{GP})$, is composed of two parts. The first part comes from K independent gamma distributions for l_k, and the second part comes from J independent Poisson distributions with parameters $\sum_k l_k \theta_{j,k}$.

$$\overbrace{\prod_k \frac{\beta_k^{\alpha_k} l_k^{\alpha_k - 1} \exp\{-\beta_k l_k\}}{\Gamma(\alpha_k)}}^{\text{likelihood of } \boldsymbol{l}} \overbrace{\prod_j \frac{\left(\sum_k l_k \theta_{j,k}\right)^{w_j} \exp\left\{-\left(\sum_k l_k \theta_{j,k}\right)\right\}}{w_j!}}^{\text{likelihood of } \boldsymbol{w} \text{ given } \boldsymbol{l}} \qquad (2)$$

4.2   The Conditional Gamma-Poisson Model

In practice, when fitting the parameters $\boldsymbol{\alpha}$ in the GP or DM model, it is often the case that the α_k go very small. Thus, in this situation, perhaps 90% of the component scores l_k are negligible, say less than 10^{-8} once normalised. Rather than maintaining these negligible values, we can allow component scores to be zero with some finite probability. The Conditional Gamma-Poisson Model, denoted CGP for short, introduces this capability. In retrospect, CGP is a sparse GP with an additional parameter per component to encourage sparsity.

The CGP model extends the Gamma-Poisson model by making the l_k zero sometimes. In the general case, the l_k are independent and zero with probability ρ_k and otherwise gamma distributed with probability $1 - \rho_k$.

$$l_k \sim \mathrm{Gamma}(\alpha_k, \beta_k) \qquad \text{for } k = 1, \ldots, K.$$

Denote the model as CGP standing for Conditional Gamma-Poisson, and the full likelihood is now $p(\boldsymbol{w}, \boldsymbol{l} \mid \boldsymbol{\beta}, \boldsymbol{\alpha}, \boldsymbol{\rho}, \boldsymbol{\Theta}, K, \mathrm{CGP})$. The full likelihood for each document, modifying the above Equation (2), replaces the term inside $\prod_k$ with

$$(1 - \rho_k) \frac{\beta_k^{\alpha_k} l_k^{\alpha_k - 1} \exp\{-\beta_k l_k\}}{\Gamma(\alpha_k)} + \rho_k 1_{l_k = 0} \qquad (3)$$

4.3   The Dirichlet-Multinomial Model

The Dirichlet-multinomial form of DCA was introduced as MPCA. In this case, the normalised latent variables m are used, and the total word count L is not modelled.

$$m \sim \text{Dirichlet}_K(\alpha) , \qquad w \sim \text{Multinomial}(L, \Theta m)$$

The first argument to the multinomial is the total count, the second argument is the vector of probabilities. Denote the model as DM, and the full likelihood is now $p(w, m \mid L, \alpha, \Theta, K, \text{DM})$. The full likelihood for each document becomes:

$$C^L_{w_1,\dots,w_J} \Gamma \left(\sum_k \alpha_k \right) \prod_k \frac{m_k^{\alpha_k - 1}}{\Gamma(\alpha_k)} \prod_j \left(\sum_k m_k \theta_{j,k} \right)^{w_j} \tag{4}$$

where C^L_w is L choose $w_1, \dots, w_J$. This model can also be derived from the Gamma-Poisson model, shown in the next section.

4.4   A Multivariate Version

Another variation of the methods is to allow grouping of the count data. Words can be grouped into separate variable sets. These groups might be "title words," "body words," and "topics" in web page analysis or "nouns," "verbs" and "adjectives" in text analysis. The groups can be treated with separate discrete distributions, as below. The J possible word types in a document are partitioned into G groups $B_1, \dots, B_G$. The total word counts for each group g is denoted $L_g = \sum_{j \in B_g} w_j$. If the vector w is split up into G vectors $w_g = \{w_j : j \in B_g\}$, and the matrix Θ is now normalised by group in each row, so $\sum_{j \in B_g} \theta_{j,k} = 1$, then a multivariate version of DCA is created so that for each group g,

$$w_g \sim \text{Multinomial} \left(L_g, \left\{ \sum_k m_k \theta_{j,k} : j \in B_g \right\} \right) .$$

Fitting and modelling methods for this variation are related to LDA or MPCA, and will not be considered in more detail here. This has the advantage that different kinds of words have their own multinomial and the distribution of different kinds is ignored. This version is demonstrated subsequently on US Senate voting records, where each multinomial is now a single vote for a particular senator.

5   Related Work

These sections begins by relating the main approaches to each other, then placing them in the context of exponential family models, and finally a brief history is recounted.

5.1   Correspondences

Various published cases of DCA can be represented in terms of this format, as given in Table 2. A multinomial with total count L and J possible outcomes is the bagged version of L discrete distributions with J possible outcomes. In the table, *NA* indicates that this aspect of the model was not required to be specified because the methodology made no use of it. Note that NMF used a cost function

Table 2. Previously Published Models

Name	Bagged	Components	$p(x/w \mid \Theta, l)$	$p(l/m)$
NMF [LS99]	yes	l	Poisson	*NA*
PLSI [Hof99]	no	m	discrete	*NA*
LDA [BNJ03]	no	m	discrete	Dirichlet
MPCA [Bun02]	yes	m	multinomial	Dirichlet
GaP [Can04]	yes	l	Poisson	gamma

formulation, and thus avoided defining likelihood models. It is shown later that its cost function corresponds to a Gamma-Poisson with parameters $\alpha = \beta = 0$ (i.e., all zero).

LDA has the multinomial of MPCA replaced by a sequence of discrete distributions, and thus the choose term drops, as per Section 3.1. PLSI is related to LDA but lacks a prior distribution on m. It does not model these latent variables using full probability theory, but instead using a weighted likelihood method [Hof99]. Thus PLSI is a non-Bayesian version of LDA, although its weighted likelihood method means it accounts for over-fitting in a principled manner.

LDA and MPCA also have a close relationship to GaP (called GP here). If the parameter α is treated as known and not estimated from the data, and the β parameter vector has the same value for each β_k, then L is *aposteriori* independent of m and Θ. In this context LDA, MPCA and GaP are equivalent models ignoring representational issues.

Lemma 1. *Given a Gamma-Poisson model of Section 4.1 where the β parameter is a constant vector with all entries the same, β, the model is equivalent to a Dirichlet-multinomial model of Section 4.3 where $m_k = l_k / \sum_k l_k$, and in addition*

$$L \sim \text{Poisson-Gamma} \left(\sum_k \alpha_k, \beta, 1 \right)$$

Proof. Consider the Gamma-Poisson model. The sum $L = \sum_j w_j$ of Poisson variables w has the distribution of a Poisson with parameter given by the sum of their means. When the sum of Poisson variables is known, the set of Poisson variables has a multinomial distribution conditioned on the sum (the total count) [Ros89]. The Poisson distributions on w then is equivalent to:

$$L \sim \text{Poisson} \left(\sum_k l_k \right) , \qquad w \sim \text{Multinomial} \left(L, \frac{1}{\sum_k l_k} \Theta l \right) .$$

Moreover, if the β parameter is constant, then $m_k = l_k / \sum_k l_k$ is distributed as Dirichlet$_K(\boldsymbol{\alpha})$, and $\sum_k l_k$ is distributed independently as a Gamma$(\sum_k \alpha_k, \beta)$. The second distribution above can then be represented as

$$\boldsymbol{w} \sim \text{Multinomial}\,(L, \boldsymbol{\Theta m})\ .$$

Note also, that marginalising out $\sum_k l_k$ convolves a Poisson and a gamma distribution to produce a Poisson-Gamma distribution for L [BS94].

If $\boldsymbol{\alpha}$ is estimated from the data in GaP, then the presence of the observed L will influence $\boldsymbol{\alpha}$, and thus the other estimates such as of $\boldsymbol{\Theta}$. In this case, LDA and MPCA will no longer be effectively equivalent to GaP. Note, Canny recommends fixing $\boldsymbol{\alpha}$ and estimating β from the data [Can04].

To complete the set of correspondences, note that in Section 7.1 it is proven that NMF corresponds to a maximum likelihood version of GaP, and thus it also corresponds to a maximum likelihood version of LDA, MPCA, and PLSI.

5.2   Notes on the Exponential Family

For the general DCA model of Section 3.2, when $p(\boldsymbol{w} \mid \boldsymbol{\Theta l})$ is in the so-called exponential family distributions [GCSR95], the expected value of $\boldsymbol{w}$ is referred to as the *dual parameter*, and it is usually the parameter we know best. For the Bernoulli with probability p, the dual parameter is p, for the Poisson with rate λ, the dual parameter is λ, and for the Gaussian with mean μ, the dual parameter is the mean. Our formulation, then, can be also be interpreted as letting $\boldsymbol{w}$ be exponential family with dual parameter given by $(\boldsymbol{\Theta l})$. Our formulation then generalises PCA in the same way that a linear model [MN89] generalises linear regression.

Note, an alternative has also been presented [CDS01] where $\boldsymbol{w}$ has an exponential family distribution with natural parameters given by $(\boldsymbol{\Theta l})$. For the Bernoulli with probability p, the natural parameter is $\log(p/(1-p))$, for the Poisson with rate λ, the natural parameter is $\log \lambda$ and for the Gaussian with mean μ, the natural parameter is the mean. This formulation generalises PCA in the same way that a generalised linear model [MN89] generalises linear regression.

5.3   Historical Notes

Several independent groups within the statistical computing and machine learning community have contributed to the development of the DCA family of methods. Some original research includes the following: grade of membership (GOM) [WM82], probabilistic latent semantic indexing (PLSI) [Hof99], non-negative matrix factorisation (NMF) [LS99], genotype inference using admixtures [PSD00], latent Dirichlet allocation (LDA) [BNJ03], and Gamma-Poisson models (GaP) [Can04]. Modifications and algorithms have also been explored as multinomial PCA (MPCA) [Bun02] and multiple aspect modelling [ML02].

The first clear enunciation of the large-scale model in its Poisson form comes from [LS99], and in its multinomial form from [Hof99] and [PSD00]. The first clear expression of the problem as a latent variable problem is given by [PSD00]. The relationship between LDA and PLSI and that NMF was a Poisson version of LDA was first pointed out by [Bun02], and proven in [GG05]. The connections to ICA come from [BJ04] and [Can04]. The general Gamma-Poisson formulation, perhaps the final generalisation to this line of work, is in [Can04].

Related techniques in the statistical community can be traced back to Latent Class Analysis developed in the 1950's, and a rich theory has since developed relating the methods to correspondence analysis and other statistical techniques [vGv99].

6   Component Assignments for Words

In standard mixture models, each document in a collection is assigned to one latent component. The DCA family of models can be interpreted as making each word in each document be assigned to one latent component. To see this, we introduce another latent vector which represents the component assignments for different words. As in Section 3.1, this can be done using a bag of components or a sequence of components representation, and no effective change occurs in the basic models, or in the algorithms so derived. What this does is expand out the term $\boldsymbol{\Theta l}$ into parts, treating it as if it is the result of marginalising out some latent variable.

We introduce a K-dimensional discrete latent vector $\boldsymbol{c}$ whose total count is L, the same as the word count. The count c_k gives the number of words in the document appearing in the k-th component. Its posterior mean makes a good diagnostic and interpretable result. A document from the sports news might have 50 "football" words, 10 "German" words, 20 "sports fan" words and 20 "general vocabulary" words.

This latent vector is derived from a larger latent matrix, $\boldsymbol{V}$ of size $J \times K$ and entries $v_{j,k}$. This has row totals w_j as given in the observed data and column totals c_k. Vectors $\boldsymbol{w}$ and $\boldsymbol{c}$ are these word appearance counts and component appearance counts, respectively, based on summing rows and columns of matrix $\boldsymbol{V}$. This is shown in Figure 3.

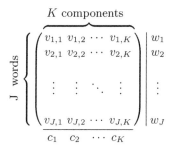

Fig. 3. A representation of a document as a contingency table

The introduction of the latent matrix V changes the forms of the likelihoods, and makes the development and analysis of algorithms easier. This section catalogues the likelihood formula, to be used when discussing algorithms.

6.1 The Gamma-Poisson Model

With the new latent matrix V, the distributions underlying the Gamma-Poisson model become:

$$l_k \sim \text{Gamma}(\alpha_k, \beta_k) \tag{5}$$

$$c_k \sim \text{Poisson}(l_k)$$

$$w_j = \sum_k v_{j,k} \qquad \text{where } v_{j,k} \sim \text{Multinomial}(c_k, \boldsymbol{\theta}_{\cdot,k}) \ .$$

The joint likelihood for a document, $p(V, l \mid \boldsymbol{\beta}, \boldsymbol{\alpha}, \boldsymbol{\Theta}, K, \text{GP})$ (the w are now derived quantities so not represented), thus becomes, after some rearrangement

$$\prod_k \frac{\beta_k^{\alpha_k} l_k^{c_k + \alpha_k - 1} \exp\{-(\beta_k + 1)l_k\}}{\Gamma(\alpha_k)} \prod_{j,k} \frac{\theta_{j,k}^{v_{j,k}}}{v_{j,k}!} \ . \tag{6}$$

Note that l can be marginalised out, yielding

$$\prod_k \frac{\Gamma(c_k + \alpha_k)}{\Gamma(\alpha_k)} \frac{\beta_k^{\alpha_k}}{(\beta_k + 1)^{c_k + \alpha_k}} \prod_{j,k} \frac{\theta_{j,k}^{v_{j,k}}}{v_{j,k}!} \tag{7}$$

and the posterior mean of l_k given c is $(c_k + \alpha_k)/(1 + \beta_k)$. Thus each $c_k \sim$ Poisson-Gamma$(\alpha_k, \beta_k, 1)$.

6.2 The Conditional Gamma-Poisson Model

The likelihood follows the GP case, except that with probability ρ_k, $l_k = 0$ and thus $c_k = 0$. The joint likelihood, $p(V, l \mid \boldsymbol{\beta}, \boldsymbol{\alpha}, \boldsymbol{\rho}, \boldsymbol{\Theta}, K, \text{CGP})$, thus becomes, after some rearrangement

$$\prod_k \left((1 - \rho_k) \left(\frac{\beta_k^{\alpha_k} l_k^{c_k + \alpha_k - 1} \exp\{-(\beta_k + 1)l_k\}}{\Gamma(\alpha_k)} \prod_j \frac{\theta_{j,k}^{v_{j,k}}}{v_{j,k}!} \right) \right. \tag{8}$$

$$\left. + \rho_k \left(1_{l_k=0} 1_{c_k=0} \prod_j 1_{v_{j,k}=0} \right) \right) \ .$$

Note that l can be marginalised out, yielding

$$\prod_k \left((1 - \rho_k) \frac{\Gamma(c_k + \alpha_k)}{\Gamma(\alpha_k)} \frac{\beta_k^{\alpha_k}}{(\beta_k + 1)^{c_k + \alpha_k}} + \rho_k 1_{c_k=0} \right) \prod_j \frac{\theta_{j,k}^{v_{j,k}}}{v_{j,k}!} \tag{9}$$

The $\theta_{j,k}$ can be pulled out under the constraint $\sum_j v_{j,k} = c_k$. The posterior mean of l_k given c is $(1 - \rho_k)(c_k + \alpha_k)/(1 + \beta_k)$.

6.3   The Dirichlet-Multinomial Model

For the Dirichlet-multinomial model, a similar reconstruction applies:

$$
\begin{aligned}
\boldsymbol{m} &\sim \text{Dirichlet}_K(\boldsymbol{\alpha}) && (10) \\
c_k &\sim \text{Multinomial}(L, \boldsymbol{m}) \\
w_j &= \sum_k v_{j,k} && \text{where } v_{j,k} \sim \text{Multinomial}(c_k, \boldsymbol{\theta}_{\cdot,k}) \ .
\end{aligned}
$$

The joint likelihood, $p(\boldsymbol{V}, \boldsymbol{m} \mid \boldsymbol{\alpha}, \boldsymbol{\Theta}, K, \text{DM})$, thus becomes, after some rearrangement

$$
L! \, \Gamma\left(\sum_k \alpha_k\right) \prod_k \frac{m_k^{c_k + \alpha_k - 1}}{\Gamma(\alpha_k)} \prod_{j,k} \frac{\theta_{j,k}^{v_{j,k}}}{v_{j,k}!} \ . \tag{11}
$$

Again, $\boldsymbol{m}$ can be marginalised out yielding

$$
L! \, \frac{\Gamma\left(\sum_k \alpha_k\right)}{\Gamma\left(L + \sum_k \alpha_k\right)} \prod_k \frac{\Gamma(c_k + \alpha_k)}{\Gamma(\alpha_k)} \prod_{j,k} \frac{\theta_{j,k}^{v_{j,k}}}{v_{j,k}!} \ . \tag{12}
$$

7   Algorithms

In developing an algorithm, the standard approach is to match an optimization algorithm to the functional form of the likelihood. When using Bayesian or some other statistical methodology, this basic approach is usually a first step, or perhaps an inner loop for some more sophisticated computational statistics.

The likelihoods do not yield easily to standard EM analysis. To see this, consider the forms of the likelihood for a single document for the GP model, and consider the probability for a latent variable $\boldsymbol{z}$ given the observed data $\boldsymbol{w}$, $p(\boldsymbol{z} \mid \boldsymbol{w}, \boldsymbol{\beta}, \boldsymbol{\alpha}, \boldsymbol{\Theta}, K, \text{GP})$. For EM analysis, one needs to be able to compute $\mathbb{E}_{\boldsymbol{z} \sim p(\boldsymbol{z} \mid \boldsymbol{w}, \boldsymbol{\Theta}, \dots)}[\log p(\boldsymbol{w}, \boldsymbol{z} \mid \boldsymbol{\Theta}, \dots)]$. There are three different forms of the likelihood seen so far depending on which latent variables $\boldsymbol{z}$ are kept on the left-hand side of the probability:

$p(\boldsymbol{w}, \boldsymbol{l} \mid \boldsymbol{\beta}, \boldsymbol{\alpha}, \boldsymbol{\Theta}, K, \textbf{GP})$:   from Equation (2) has the term $\left(\sum_k l_k \theta_{j,k}\right)^{w_j}$, which means there is no known simple posterior distribution for $\boldsymbol{l}$ given $\boldsymbol{w}$.

$p(\boldsymbol{w}, \boldsymbol{l}, \boldsymbol{V} \mid \boldsymbol{\beta}, \boldsymbol{\alpha}, \boldsymbol{\Theta}, K, \textbf{GP})$:   from Equation (6) has the term $l_k^{c_k + \alpha_k - 1}$ which links the two latent variables $\boldsymbol{l}$ and $\boldsymbol{V}$, and prevents a simple evaluation of $\mathbb{E}_{\boldsymbol{l}, \boldsymbol{V}}[v_{j,k}]$ as required for the expected log probability.

$p(\boldsymbol{w}, \boldsymbol{V} \mid \boldsymbol{\beta}, \boldsymbol{\alpha}, \boldsymbol{\Theta}, K, \textbf{GP})$:   from Equation (7) has the term $\Gamma(c_k + \alpha_k)$ (where $c_k = \sum_j v_j, k$), which means there is no known simple posterior distribution for $\boldsymbol{V}$ given $\boldsymbol{w}$.

Now one could always produce an EM-like algorithm by separately updating $\boldsymbol{l}$ and $\boldsymbol{V}$ in turn according to some mean formula, but the guarantee of convergence of $\boldsymbol{\Theta}$ to a maximum posterior or likelihood value will not apply. In this spirit earlier authors point out that EM-like principles apply and use EM terminology since EM methods would apply if l was observed[6]. For the exponential family, which this problem is in, the variational approximation algorithm with Kullback-Leibler divergence corresponds to an extension of the EM algorithm [GB00, Bun02]. This variational approach is covered below.

Algorithms for this problem follow some general approaches in the statistical computing community. Three basic approaches are presented here: a variational approximation, Gibbs sampling, and Rao-Blackwellised Gibbs sampling. A maximum likelihood algorithm is not presented because it can be viewed as a simplification of the algorithms here.

7.1 Variational Approximation with Kullback-Leibler Divergence

This approximate method was first applied to the sequential variant of the Dirichlet-multinomial version of the problem by [BNJ03]. A fuller treatment of these variational methods for the exponential family is given in [GB00, Bun02].

In this approach a factored posterior approximation is made for the latent variables:

$$p(l, V \mid w, \beta, \alpha, \Theta, K, \text{GP}) \approx q(l, V) = q_l(l) q_V(V)$$

and this approximation is used to find expectations as part of an optimization step. The EM algorithm results if an equality holds. The functional form of the approximation can be derived by inspection of the recursive functional forms (see [Bun02] Equation (4)):

$$q_l(l) \propto \exp\left(\mathbb{E}_{V \sim q_V(V)}\left[\log p\left(l, V, w \mid \Theta, \alpha, \beta, K\right)\right]\right) \qquad (13)$$
$$q_V(V) \propto \exp\left(\mathbb{E}_{l \sim q_l(l)}\left[\log p\left(l, v, w \mid \Theta, \alpha, \beta, K\right)\right]\right) \ .$$

An important computation used during convergence in this approach is a lower bound on the individual document log probabilities. This naturally falls out during computation (see [Bun02] Equation (6)). Using the approximation $q(l, V)$ defined by the above proportions, the bound is given by

$$\log p\left(w \mid \Theta, \alpha, \beta, K\right)$$
$$\geq \quad \mathbb{E}_{l, V \sim q(l, V)}\left[\log p\left(l, V, w \mid \Theta, \alpha, \beta, K\right)\right] + I(q_l(l)) + I(q_V(V)) \ .$$

The variational approximation applies to the Gamma-Poisson version and the Dirichlet-multinomial version.

For the Gamma-Poisson Model: Looking at the recursive functionals of Equation (13) and the likelihood of Equation (6), it follows that $q_l()$ must be K independent Gammas one for each component, and $q_V()$ must be J independent multinomials, one for each word. The most general case for the approximation $q()$ is thus

---

[6] The likelihood $p(w, V \mid l, \beta, \alpha, \Theta, K, \text{GP})$ can be treated with EM methods using the latent variable V and leaving l as if it was observed.

$$l_k \sim \text{Gamma}(a_k, b_k)$$
$$\{v_{j,k} : k = 1, \ldots, K\} \sim \text{Multinomial}(w_j, \{n_{j,k} : k = 1, \ldots, K\}) ,$$

which uses approximation parameters (a_k, b_k) for each Gamma and and $\mathbf{n}_{.,k}$ (normalised as $\sum_k n_{j,k} = 1$) for each multinomial. These parameters form two vectors $\mathbf{a}, \mathbf{b}$ and a matrix $\mathbf{N}$ respectively. The approximate posterior takes the form $q_l(\mathbf{l} \mid \mathbf{a}, \mathbf{b}) q_V(\mathbf{V} \mid \mathbf{N})$.

Using these approximating distributions, and again looking at the recursive functionals of Equation (13), one can extract the rewrite rules for the parameters:

$$n_{j,k} = \frac{1}{Z_j} \theta_{j,k} \exp\left(\mathbb{E}\left[\log l_k\right]\right), \tag{14}$$

$$a_k = \alpha_k + \sum_j w_j n_{j,k},$$

$$b_k = 1 + \beta_k,$$

where $\mathbb{E}\left[\log l_k\right] \equiv \mathbb{E}_{l_k \sim p(l_k \mid a_k, b_k)}\left[\log l_k\right] = \Psi_0(a_k) - \log b_k,$

$$Z_j \equiv \sum_k \theta_{j,k} \exp\left(\mathbb{E}\left[\log l_k\right]\right).$$

Here, $\Psi_0()$ is the digamma function, defined as $\frac{d \ln \Gamma(x)}{dx}$ and available in most scientific libraries. These equations form the first step of each major cycle, and are performed on each document.

The second step is to re-estimate the model parameters $\boldsymbol{\Theta}$ using the posterior approximation by maximising the expectation of the log of the full posterior probability

$$\mathbb{E}_{l, V \sim q_l(l) q_V(V)}\left[\log p\left(\mathbf{l}, \mathbf{V}, \mathbf{w}, \boldsymbol{\Theta} \mid \boldsymbol{\alpha}, \boldsymbol{\beta}, K\right)\right] .$$

This incorporates Equation (6) for each document, and a prior for each k-th column of $\boldsymbol{\Theta}$ of Dirichlet$_J(\boldsymbol{\gamma})$ (the last model item in Section 4.1). Denote the intermediate variables $n_{j,k}$ for the i-th document by adding a (i) subscript, as $n_{j,k,(i)}$, and likewise for $w_{j,(i)}$. All these log probability formulas yield linear terms in $\theta_{j,k}$, thus with the normalising constraints for $\boldsymbol{\Theta}$ one gets

$$\theta_{j,k} \propto \sum_i w_{j,(i)} n_{j,k,(i)} + \gamma_j . \tag{15}$$

The lower bound on the log probability of Equation (14), after some simplification and use of the rewrites of Equation (14), becomes

$$\log \frac{1}{\prod_j w_j!} - \sum_k \log \frac{\Gamma(\alpha_k) b_k^{a_k}}{\Gamma(a_k) \beta_k^{\alpha_k}} + \sum_k (\alpha_k - a_k)\mathbb{E}\left[\log l_k\right] + \sum_j w_j \log Z_j . \tag{16}$$

The variational approximation algorithm for the Gamma-Poisson version is summarised in Figure 4. An equivalent algorithm is produced if words are presented sequentially instead of being bagged.

1. Initialise a for each document. The uniform initialisation would be $a_k = \left(\sum_k \alpha_k + L \right) / K$. Note N is not stored.
2. Do for each document:
   (a) Using Equations (14), recompute N and update a in place.
   (b) Concurrently, compute the log-probability bound of Equation (16), and add to a running total.
   (c) Concurrently, maintain the sufficient statistics for Θ, the total $\sum_i w_{j,(i)} n_{j,k,(i)}$ for each j, k over documents.
   (d) Store a for the next cycle and discard N.
3. Update Θ using Equation (15), normalising appropriately.
4. Report the total log-probability bound, and repeat, starting at Step 2.

Fig. 4. K-L Variational Algorithm for Gamma-Poisson

Complexity: Because Step 2(a) only uses words appearing in a document, the full Step 2 is $O(SK)$ in time complexity where S is the number of words in the full collection. Step 3 is $O(JK)$ in time complexity. Space complexity is $O(IK)$ to store the intermediate parameters a for each document, and the $O(2JK)$ to store Θ and its statistics. In implementation, Step 2 for each document is often quite slow, and thus both a and the document word data can be stored on disk and streamed, thus the main memory complexity is $O(2JK)$ since the $O(S)$ and $O(IK)$ terms are on disk. If documents are very small (e.g., $S/I \ll K$, for instance "documents" are sentences or phrases), then this does not apply.

Correspondence with NMF: A precursor to the GaP model is non-negative matrix factorisation (NMF) [LS99], which is based on the matrix approximation paradigm using Kullback-Leibler divergence. The algorithm itself, converted to the notation used here, is as follows

$$l_{k,(i)} \longleftarrow l_{k,(i)} \sum_j \frac{\theta_{j,k}}{\sum_j \theta_{j,k}} \frac{w_{j,(i)}}{\sum_k \theta_{j,k} l_{k,(i)}} \qquad \theta_{j,k} \longleftarrow \theta_{j,k} \sum_i \frac{l_{k,(i)}}{\sum_i l_{k,(i)}} \frac{w_{j,(i)}}{\sum_k \theta_{j,k} l_{k,(i)}}$$

Notice that the solution is indeterminate up to a factor ψ_k. Multiply $l_{k,(i)}$ by ψ_k and divide $\theta_{j,k}$ by ψ_k and the solution still holds. Thus, without loss of generality, let $\theta_{j,k}$ be normalised on j, so that $\sum_j \theta_{j,k} = 1$.

Lemma 2. *The NMF equations above, where Θ is returned normalised, occur at a maxima w.r.t. Θ and l for the Gamma-Poisson likelihood $\prod_i p(w_{(i)} \mid \Theta, l_{(i)}, \alpha = 0, \beta = 0, K, GP)$.*

Proof. To see this, the following will be proven. Take a solution to the NMF equations, and divide $\theta_{j,k}$ by a factor $\psi_k = \sum_j \theta_{j,k}$, and multiply $l_{k,(i)}$ by the same factor. This is equivalent to a solution for the following rewrite rules

$$l_{k,(i)} \longleftarrow l_{k,(i)} \sum_j \theta_{j,k} \frac{w_{j,(i)}}{\sum_k \theta_{j,k} l_{k,(i)}} \qquad \theta_{j,k} \propto \theta_{j,k} \sum_i l_{k,(i)} \frac{w_{j,(i)}}{\sum_k \theta_{j,k} l_{k,(i)}}$$

where $\theta_{j,k}$ is kept normalised on j. These equations hold at a maxima to the likelihood $\prod_i p(w_{(i)} \mid \Theta, l_{(i)}, \alpha = 0, \beta = 0, K, \mathrm{GP})$. The left equation corresponds to a maxima w.r.t. $l_{(i)}$ (note the Hessian for this is easily shown to be negative indefinite), and the right is the EM equations for the likelihood. w.r.t. $\boldsymbol{\Theta}$.

To show equivalence of the above and the NMF equations, first prove the forward direction. Take the scaled solution to NMF. The NMF equation for $l_{k,(i)}$ is equivalent to the equation for $l_{k,(i)}$ in the lemma. Take the NMF equation for $\theta_{j,k}$ and separately normalise both sides. The $\sum_i l_{k,(i)}$ term drops out and one is left with the equation for $\theta_{j,k}$ in the lemma. Now prove the backward direction. It is sufficient to show that the NMF equations hold for the solution to the rewrite rules in the lemma, since $\theta_{j,k}$ is already normalised. The NMF equation for $l_{k,(i)}$ clearly holds. Assuming the rewrite rules in the lemma hold, then

$$
\begin{aligned}
\theta_{j,k} &= \frac{\theta_{j,k} \sum_i \left(l_{k,(i)} w_{j,(i)} \big/ \sum_k \theta_{j,k} l_{k,(i)} \right)}{\sum_j \theta_{j,k} \sum_i \left(l_{k,(i)} w_{j,(i)} \big/ \sum_k \theta_{j,k} l_{k,(i)} \right)} \\[2ex]
&= \frac{\theta_{j,k} \sum_i \left(l_{k,(i)} w_{j,(i)} \big/ \sum_k \theta_{j,k} l_{k,(i)} \right)}{\sum_i l_{k,(i)} \sum_j \left(\theta_{j,k} w_{j,(i)} \big/ \sum_k \theta_{j,k} l_{k,(i)} \right)} \qquad \text{(reorder sum)} \\[2ex]
&= \frac{\theta_{j,k} \sum_i \left(l_{k,(i)} w_{j,(i)} \big/ \sum_k \theta_{j,k} l_{k,(i)} \right)}{\sum_i l_{k,(i)}} \qquad \text{(apply first rewrite rule)}
\end{aligned}
$$

Thus the second equation for NMF holds.

Note, including a latent variable such as $\boldsymbol{l}$ in the likelihood (and not dealing with it using EM methods) does not achieve a correct maximum likelihood solution for the expression $\prod_i p(\boldsymbol{w}_{(i)} \,|\, \boldsymbol{\Theta}, \boldsymbol{\alpha} = 0, \boldsymbol{\beta} = 0, K, \text{GP})$. In practice, this is a common approximate method for handling latent variable problems, and can lead more readily to over-fitting.

For the Dirichlet-Multinomial Model: The variational approximation takes a related form. The approximate posterior is given by:

$$
\boldsymbol{m} \sim \text{Dirichlet}(\boldsymbol{a})
$$
$$
\{v_{j,k} : k = 1, \dots, K\} \sim \text{multinomial}(w_j, \{n_{j,k} : k = 1, \dots, K\})
$$

This yields the same style update equations as Equations (14) except that $\beta_k = 1$

$$
n_{j,k} = \frac{1}{Z_j} \theta_{j,k} \exp\left(\mathbb{E}\left[\log m_k \right] \right) , \tag{17}
$$
$$
a_k = \alpha_k + \sum_j w_j n_{j,k} ,
$$

where $\mathbb{E}\left[\log m_k \right] \equiv \mathbb{E}_{m_k \sim p(m_k \,|\, \boldsymbol{a})}\left[\log m_k \right] = \Psi_0(a_k) - \Psi_0\left(\sum_k a_k \right)$,

$$
Z_j \equiv \sum_k \theta_{j,k} \exp\left(\mathbb{E}\left[\log m_k \right] \right) .
$$

Equation (15) is also the same. The lower bound on the individual document log probabilities, $\log p\left(\boldsymbol{w} \mid \boldsymbol{\Theta}, \boldsymbol{\alpha}, K, \mathrm{DM}\right)$ now takes the form

$$\log\left(C_{\boldsymbol{w}}^L\right) - \log \frac{\Gamma\left(\sum_k a_k\right) \prod_k \Gamma(\alpha_k)}{\Gamma\left(\sum_k \alpha_k\right) \prod_k \Gamma(a_k)} + \sum_k (\alpha_k - a_k) \mathbb{E}\left[\log m_k\right] + \sum_j w_j \log Z_j \ . \tag{18}$$

The correspondence with Equation (16) is readily seen.

The algorithm for Dirichlet-multinomial version is related to that in Figure 4. Equations (17) replace Equations (14), Equation (18) replaces Equation (16), and the initialisation for a_k should be 0.5, a Jeffreys prior.

7.2   Direct Gibbs Sampling

There are two styles of Gibbs sampling that apply to DCA. The first is a basic Gibbs sampling first proposed by Pritchard, Stephens and Donnelly [PSD00]. Gibbs sampling is a conceptually simple method. Each unobserved variable in the problem is resampled in turn according to its conditional distribution. We compute its posterior distribution conditioned on all other variables, and then sample a new value for the variable using the posterior. For instance, an ordering we might use in this problem is: $\boldsymbol{l}_{(1)}, \boldsymbol{V}_{(1)}, \boldsymbol{l}_{(2)}, \boldsymbol{V}_{(2)}, \ldots, \boldsymbol{l}_{(I)}, \boldsymbol{V}_{(I)}, \boldsymbol{\Theta}$. All the low level sampling in this section use well known distributions such as gamma or multinomial, and are available in standard scientific libraries.

To develop this approach for the Gamma-Poisson, look at the full posterior, which is a product of individual document likelihoods with the prior for $\boldsymbol{\Theta}$ from the last model item in Section 4.1. The constant terms have been dropped.

$$\prod_i \left(\prod_k \frac{\beta_k^{\alpha_k} l_{k,(i)}^{c_{k,(i)}+\alpha_k-1} \exp\{-(\beta_k+1)l_{k,(i)}\}}{\Gamma(\alpha_k)} \prod_{j,k} \frac{\theta_{j,k}^{v_{j,k,(i)}}}{v_{j,k,(i)}!} \right) \prod_{j,k} \theta_{j,k}^{\gamma_j} \tag{19}$$

Each of the conditional distributions used in the Gibbs sampling are proportional to this. The first conditional distribution is $p(\boldsymbol{l}_{(i)} \mid \boldsymbol{V}_{(i)}, \boldsymbol{\beta}, \boldsymbol{\alpha}, \boldsymbol{\Theta}, K, \mathrm{GP})$. From this, isolating the terms just in $\boldsymbol{l}_{(i)}$, we see that each $l_{k,(i)}$ is conditionally gamma distributed. Likewise, each $\boldsymbol{v}_{\cdot,k,(i)}$ is multinomial distributed given $\boldsymbol{l}_{(i)}$ and $\boldsymbol{\Theta}$, and each $\boldsymbol{\theta}_{\cdot,k}$ is Dirichlet distributed given all the $\boldsymbol{l}_{(i)}$ and $\boldsymbol{V}_{(i)}$ for each i. The other models are similar. An additional effort is required to arrange the parameters and sequencing for efficient use of memory.

The major differentiator for Gibbs sampling is the resampling of the latent component vector $\boldsymbol{l}$. The sampling schemes used for each version are given in Table 3. Some care is required with the conditional Gamma-Poisson. When $c_k = 0$, the sampling for l_k needs to decide whether to use the zero case or the non-zero case. This uses Equation (9) to make the decision, and then resorts to Equation (8) if it is non-zero.

The direct Gibbs algorithm for the general case is given in Figure 5. This Gibbs scheme turns out to correspond to the variational approximation, excepting that sampling is done instead of maximisation or expectation.

Table 3. Sampling components for direct Gibbs on a single document

Model	Sampling
GP	$l_k \sim \text{Gamma}(c_k + \alpha_k, 1 + \beta_k)$.
CGP	If $c_k = 0$, then Conditional Gamma-Poisson with rate $\frac{p_k(1+\beta_k)^{\alpha_k}}{(1-p_k)\beta_k^{\alpha_k}+p_k(1+\beta_k)^{\alpha_k}}$ and $\text{Gamma}(\alpha_k, 1+\beta_k)$. If $c_k \neq 0$, revert to the above Gamma-Poisson case.
DM	$\boldsymbol{m} \sim \text{Dirichlet}(\{c_k + \alpha_k : k\})$.

1. For each document i, retrieve the last $\boldsymbol{c}_{(i)}$ from store, then
   (a) Sample the latent component variables $\boldsymbol{l}_{(i)}$ (or its normalised counterpart $\boldsymbol{m}_{(i)}$) as per Table 3.
   (b) For each word j in the document with positive count $w_{j,(i)}$, the component counts vector, from Equation (5) and Equation (10),

   $$\{v_{j,k,(i)} : k = 1, \ldots, K\} \sim \text{Multinomial}\left(w_{j,(i)}, \left\{\frac{l_{k,(i)}\theta_{j,k}}{\sum_k l_{k,(i)}\theta_{j,k}} : k\right\}\right) .$$

   Alternatively, if the sequence-of-components version is to be used, the component for each word can be sampled in turn using the corresponding Bernoulli distribution.
   (c) Concurrently, accumulate the log-probability $p\left(\boldsymbol{w}_{(i)} \mid \boldsymbol{l}_{(i)}, \boldsymbol{\alpha}, \boldsymbol{\beta}, \boldsymbol{\Theta}, K, \text{GP}\right)$, $p\left(\boldsymbol{w}_{(i)} \mid \boldsymbol{l}_{(i)}, \boldsymbol{\alpha}, \boldsymbol{\beta}, \boldsymbol{\rho}, \boldsymbol{\Theta}, K, \text{CGP}\right)$, or $p\left(\boldsymbol{w}_{(i)} \mid \boldsymbol{m}_{(i)}, L_{(i)}, \boldsymbol{\alpha}, \boldsymbol{\beta}, \boldsymbol{\Theta}, K, \text{DM}\right)$.
   (d) Concurrently, maintain the sufficient statistics for $\boldsymbol{\Theta}$, the total $\sum_i v_{j,k,(i)}$ for each j, k over documents.
   (e) Store $\boldsymbol{c}_{(i)}$ for the next cycle and discard $\boldsymbol{V}_{(i)}$.
2. Using a Dirichlet prior for rows of $\boldsymbol{\Theta}$, and having accumulated all the counts $\boldsymbol{V}_{(i)}$ for each document in sufficient statistics for $\boldsymbol{\Theta}$, then its posterior has rows that are Dirichlet. Sample.
3. Report the total log-probability, and report.

Fig. 5. One Major Cycle of Gibbs Algorithm for DCA

The log probability of the words $\boldsymbol{w}$ can also accumulated in step 1(c). While they are in terms of the latent variables, they still represent a reasonably unbiased estimate of the likelihoods such as $p\left(\boldsymbol{w}_{(1)}, \ldots, \boldsymbol{w}_{(I)} \mid \boldsymbol{\alpha}, \boldsymbol{\beta}, \boldsymbol{\Theta}, K, \text{GP}\right)$.

7.3 Rao-Blackwellised Gibbs Sampling

Rao-Blackwellisation of Gibbs sampling [CR96] combines closed form updates of variables with Gibbs sampling. It does so by a process called marginalisation or variable elimination. When feasible, it can lead to significant improvements, the general case for DCA. Griffiths and Steyvers [GS04] introduced this algorithm for LDA, and it easily extends to the Gamma-Poisson model and its conditional variant with little change to the sampling routines.

When using this approach, the first step is to consider the full posterior probability and see which variables can be marginalised out without introducing computational complexity in the sampling. For the GP model, look at the posterior given in Equation (19). Equations (7) shows that the $l_{(i)}$'s can be marginalised out. Likewise, Θ can be marginalised out because it is an instance of a Dirichlet. This yields a Gamma-Poisson posterior $p\left(V_{(1)}, \ldots, V_{(I)} \mid \alpha, \beta, K, \mathrm{GP}\right)$, with constants dropped:

$$\prod_i \left(\prod_k \frac{\Gamma(c_{k,(i)} + \alpha_k)}{(1 + \beta_k)^{c_{k,(i)} + \alpha_k}} \prod_{j,k} \frac{1}{v_{j,k,(i)}!} \right) \prod_k \frac{\prod_j \Gamma\left(\gamma_j + \sum_i v_{j,k,(i)}\right)}{\Gamma\left(\sum_j \gamma_j + \sum_i c_{k,(i)}\right)} \tag{20}$$

Below it is shown that a short sampling routine can be based on this.

A similar formula applies in the conditional GP case using Equation (9) for the marginalisation of $l_{(i)}$'s. The first term with $\prod_k$ in Equation (20) becomes

$$(1 - \rho_k) \frac{\Gamma(c_{k,(i)} + \alpha_k)}{\Gamma(\alpha_k)} \frac{\beta_k^{\alpha_k}}{(1 + \beta_k)^{c_{k,(i)} + \alpha_k}} + \rho_k 1_{c_{k,(i)} = 0} .$$

Likewise a similar formula applies in the Dirichlet-multinomial version using Equation (12):

$$\prod_i \left(\prod_k \Gamma(c_{k,(i)} + \alpha_k) \prod_{jk} \frac{1}{v_{j,k,(i)}!} \right) \prod_k \frac{\prod_j \Gamma\left(\gamma_j + \sum_i v_{j,k,(i)}\right)}{\Gamma\left(\sum_j \gamma_j + \sum_i c_{k,(i)}\right)} \tag{21}$$

Here a term of the form $\Gamma\left(\sum_k (c_{k,(i)} + \alpha_k)\right)$ drops out because $\sum_k c_{k,(i)} = L_{(i)}$ is known and thus constant.

Now the posterior distributions have been marginalised for each of the three models, GP, CGP and DM, a Gibbs sampling scheme needs to be developed. Each set $\{v_{j,k,(i)} : k \in 1, \ldots, K\}$ sums to $w_{j,(i)}$, moreover the forms of the functions in Equations (20) and (21) are quite nasty. A way out of this mess is to convert the scheme from a bag of words model, implicit in the use of $V_{(i)}$ and $w_{(i)}$, to a sequence of words model.

This proceeds as follows. Run along the $L_{(i)}$ words in a document and update the corresponding component assignment for each word. Component assignments for the i-th document are in a $L_{(i)}$-dimensional vector $k_{(i)}$, where each entry takes a value from $1, \ldots, K$. Suppose the l-th word has word index j_l. In one step, change the counts $\{v_{j_l,k,(i)} : k \in 1, \ldots, K\}$ by one (one is increased and one is decreased) keeping the total $w_{j_l,(i)}$ constant. For instance, if a word is originally in component k_1 but updating by Gibbs sampling to k_2, then decrease $v_{j_l,k_1,(i)}$ by one and increase $v_{j_l,k_2,(i)}$ by one. Do this for $L_{(i)}$ words in the document, for each document. Thus at word l for the i-th document, we sample component assignment $k_{l,(i)}$ according to the posterior for $k_{l,(i)}$ with all other assignments fixed. This posterior is proportional to (the denominator is a convenient constant)

$$\frac{p\left(V \mid \text{sequential}, \alpha, \beta, K, \mathrm{GP}\right)\big|_{v_{j_l,k,(i)} \leftarrow v_{j_l,k,(i)} + 1_{k \neq k_l}}}{p\left(V \mid \text{sequential}, \alpha, \beta, K, \mathrm{GP}\right)\big|_{v_{j_l,k_l,(i)} \leftarrow v_{j_l,k_l,(i)} - 1}} ,$$

where the notation "sequential" is added to the right-hand side because the combinatoric terms $v_{j,k,(i)}!$ of Equation (20) need to be dropped. This formula simplifies dramatically because $\Gamma(x+1)/\Gamma(x) = x$.

Derived sampling schemes are given in Table 3. The (i) subscript is dropped and assumed for all counts, and $j = j_l$ is the word index for the word whose component index is being resampled. Since k_l is being sampled, a K dimensional probability vector is needed. The table gives the unnormalised form.

Table 4. Sampling $k_l = k$ given $j = j_l$ for Rao-Blackwellised Gibbs

Model	Sampling Proportionality
GP	$\dfrac{\gamma_j + \sum_i v_{j,k}}{\sum_j \gamma_j + \sum_i c_k} \dfrac{c_k + \alpha_k}{1 + \beta_k}$
CGP	When $c_k > 0$ use the proportionality of the GP case, and otherwise $\dfrac{\gamma_j + \sum_i v_{j,k}}{\sum_j \gamma_j + \sum_i c_k} \dfrac{\alpha_k}{1 + \beta_k} \dfrac{(1-\rho_k)\beta_k^{\alpha_k}}{(1-\rho_k)\beta_k^{\alpha_k} + \rho_k(1+\beta_k)^{\alpha_k}}$
DM	$\dfrac{\gamma_j + \sum_i v_{j,k}}{\sum_j \gamma_j + \sum_i c_k}(c_k + \alpha_k).$

---

1. Maintain the sufficient statistics for $\boldsymbol{\Theta}$, given by $\sum_i v_{j,k,(i)}$ for each j and k, and the sufficient statistics for the component proportions $\boldsymbol{l}_{(i)}/\boldsymbol{m}_{(i)}$ given by $\boldsymbol{c}_{(i)}$.
2. For each document i, retrieve the $L_{(i)}$ component assignments for each word then:
    (a) Recompute statistics for $\boldsymbol{l}_{(i)}/\boldsymbol{m}_{(i)}$ given by $c_{k,(i)} = \sum_j v_{j,k,(i)}$ for each k from the individual component assignment for each word.
    (b) For each word l with word index j_l and component assignment k_l in the document, resample the component assignment for this word according to the marginalised likelihoods in this section.
        i. First decrement $v_{j_l,k_l,(i)}$ and $c_{k_l,(i)}$ by one to remove the component assignment for the word.
        ii. Sample $k_l = k$ proportionally as in Table 4.
        iii. Increment $v_{j_l,k_l,(i)}$ and $c_{k_l,(i)}$.
    (c) Concurrently, record the log-probability such as $p\left(\boldsymbol{w}_{(i)} \mid \boldsymbol{V}_{(i)}, \boldsymbol{\alpha}, \boldsymbol{\beta}, \boldsymbol{\Theta}, K, \mathrm{GP}\right)$ for the appropriate model.
    (d) Concurrently, update the sufficient statistics for $\boldsymbol{l}_{(i)}/\boldsymbol{m}_{(i)}$ and $\boldsymbol{\Theta}$.

Fig. 6. One Major Cycle of Rao-Blackwellised Gibbs Algorithm for DCA

This Rao-Blackwellised Gibbs algorithm is given in Figure 6. As before, an approximately unbiased log probability can be recorded in Step 2(c). This requires a value for $\boldsymbol{\Theta}$. While the sufficient statistics could be used to supply the current mean estimate for $\boldsymbol{\Theta}$, this is not a true sampled quantity. An alternative method is to make a sample of $\boldsymbol{\Theta}$ in each major cycle and use this.

Implementation Notes: Due to Rao-Blackwellisation, both the $l_{(i)}$'s and Θ are effectively re-estimated with each sampling step, instead of once after the full pass over documents. This is most effective during early stages, and explains the superiority of the method observed in practice. Moreover, it means only one storage slot for Θ is needed (to store the sufficient statistics), whereas in direct Gibbs two slots are needed (current value plus the sufficient statistics). This represents a major saving in memory. Finally, the $l_{(i)}$'s and Θ can be sampled at any stage of this process (because their sufficient statistics make up the totals appearing in the formula), thus Gibbs estimates for them can be made as well during the MCMC process.

7.4   Historical Notes

Some previous algorithms can now be placed into context.

NMF: Straight maximum likelihood, e.g. in [LS99], expressed in terms of Kullback-Leibler divergence minimization, where optimisation jointly applies to the latent variables (see Section 7.1).

PLSI: Annealed maximum likelihood [Hof99], best viewed in terms of its clustering precursor such as by [HB97],

Various Gibbs: Gibbs sampling on $V_{(i)}$, $l_{(i)}/m_{(i)}$ and Θ in turn using a full probability distribution by [PSD00], or Gibbs sampling on $V_{(i)}$ alone (or equivalently, component assignments for words in the sequence of words representation) after marginalising out $l_{(i)}/m_{(i)}$ and Θ by [GS04],

LDA: variational approximation with Kullback-Leibler divergence by [BNJ03], a significant introduction because of its speed.

Expectation propagation [ML02] requires $O(KS)$ latent variables stored, a prohibitive expense compared to the $O(S)$ or $O(KI)$ of other algorithms. Thus it has not been covered here.

7.5   Other Aspects for Estimation and Use

A number of other algorithms are needed to put these models into regular use.

Component parameters: The treatment so far has assumed the parameter vectors α and β are given. It is more usual to estimate these parameters with the rest of the estimation tasks as done by [BNJ03, Can04]. This is feasible because the parameters are shared across all the data, unlike the component vectors themselves.

Estimating the number of components K: The number of components K is usually a constant assumed a priori. But it may be helpful to treat as a fittable parameter or a random variable that adapts to the data. In popular terms, this could be used to find the "right" number of components, though in practice and theory such a thing might not exist. To obtain best-fitting K, we can employ cross-validation, or we assess the *evidence* (or marginal likelihood) for the model given a particular choice of K [CC95, BJ04]. In particular, evidence is the posterior probability of the data given the choice of K after all other parameters have been integrated out.

Use on new data: A typical use of the model requires performing inference related to a particular document. Suppose, for instance, one wished to estimate how well a snippet of text, a query, matches a document. Our document's components are summarised by the latent variables m (or l). If the new query is represented by q, then $p(q|m, \Theta, K, \mathrm{GP})$ is the matching quantity one would like ideally. Since m is unknown, we must average over it. Various methods have been proposed [ML02, BJ04].

Alternative components: Hierarchical components have been suggested [BJ04] as a way of organising an otherwise large flat component space. For instance, the Wikipedia with over half a million documents can easily support the discovery of several hundred components. Dirichlet processes have been developed as an alternative to the K-dimensional component priors in the Dirichlet-multinomial/ discrete model [YYT05], although in implementation the effect is to use K-dimensional Dirichlets for a large K and delete low performing components.

8 Applications

This section briefly discusses two applications of the methods.

8.1 Voting Data

One type of political science data are the *roll calls*. There were 459 roll calls in the US Senate in the year 2003. For each of those, the vote of every senator was recorded in three ways: 'Yea', 'Nay' and 'Not Voting'. The outcome of the roll call can be positive (e.g., Bill Passed, Nomination Confirmed) corresponding to 'Yea', or negative (e.g., Resolution Rejected, Veto Sustained). Hence, the outcome of the vote can be interpreted as the 101st senator, by associating positive outcomes with 'Yea' and negative outcomes with 'Nay'.

Application of the Method: We can now map the roll call data to the DCA framework. For each senator X we form two 'words', where $w_{X,y}$ implies that X voted 'Yea', and $w_{X,n}$ implies that X voted 'Nay'. Each roll call can be interpreted as a document containing a single occurrence of some of the available words. The pair of words $w_{X,y}, w_{X,n}$ is then treated as a binomial, so the multivariate formulation of Section 4.4 is used. Priors for Θ were Jeffreys priors, α was (0.1,0.1,...,0.1), and regular Gibbs sampling was used.

Special-purpose models are normally used for interpreting roll call data in political science, and they often postulate a model of rational decision making. Each senator is modelled as a position or an *ideal point* in a continuous spatial model of preferences [CJR04]. For example, the first dimension often delineates the liberal-conservative preference, and the second region or social issues preference. The proximities between ideal points 'explain' the positive correlations between the senators' votes. The ideal points for each senator can be obtained either by optimization, for instance, with the optimal classification algorithm [Poo00], or through Bayesian modelling [CJR04].

Unlike the spatial models, the DCA interprets the correlations between votes through membership of the senators in similar blocs. Blocs correspond to latent component variables. Of course, we can speak only of the probability that a particular senator is a member of a particular bloc. The corresponding probability vector is normalized and thus assures that a senator is always a member of one bloc on the average. The outcome of the vote is also a member of several blocs, and we can interpret the membership as a measure of how influential a particular bloc is.

Our latent senator (bloc) can be seen as casting votes in each roll call. We model the behavior of such latent blocs across the roll calls, and record it: it has a behavior of its own. In turn, we also model the membership of each senator to a particular bloc, which is assumed to be constant across all the blocs.

A related family of approaches is based on modelling relations or networks using blocks or groups. There, a roll call would be described by one network, individual senators would be nodes in that network, and a pair of nodes is connected if the two senators agreed. Discrete latent variables try to explain the existence of links between entities in terms of senators' membership to blocks, e.g., [HLL83, SN97].

Several authors prefer the block-model approach to modelling roll call data [WMM05]. The membership of senators to the same block along with a high probability for within-block agreements will explain the agreements between senators. While a bloc can be seen as having an opinion about each issue, a block does not (at least not explicitly). The authors also extended this model to 'topics', where the membership of senator to a particular block depends on the topic of the issue; namely, the agreement between senators depends on what is being discussed. The topic is also associated with the words that appear in the description of an issue.

Visualization: We can analyze two aspects of the DCA model as applied to the roll call data: we can examine the membership of senators in blocs, and we can examine the actions of blocs for individual issues. The approach to visualization is very similar, as we are visualizing a set of probability vectors. We can use the gray scale to mirror the probabilities ranging from 0 (white) to 1 (black).

As yet, we have not mentioned the choice of K - the number of blocs. Although the number of blocs can be a nuisance variable, such a model is distinctly more difficult to show than one for a fixed K. We obtain the following negative logarithms to the base 2 of the model's likelihood for $K = 4, 5, 6, 7, 10$: 9448.6406, 9245.8770, 9283.1475, 9277.0723, 9346.6973. We see that $K = 5$ is overwhelmingly selected over all others, with $K = 4$ being far worse. This means that with our model, we best describe the roll call votes with the existence of five blocs. Fewer blocs do not capture the nuances as well, while more blocs would not yield reliable probability estimates given such an amount of data. Still, those models are also valid to some extent. It is just that for a single visualization we pick the best individual one of them.

We will now illustrate the membership of senators in blocs. Each senator is represented with a vertical bar of 5 squares that indicate his or her membership

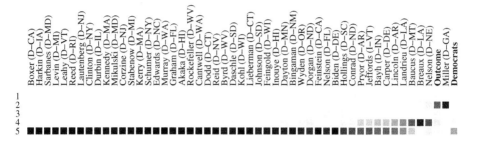

Fig. 7. Component membership for Democrats

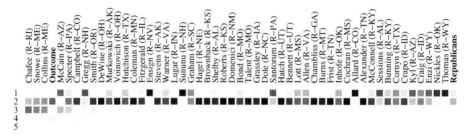

Fig. 8. Component membership for Republicans

in blocs. We have arranged the senators from left to right using the binary PCA approach of [deL03]. This ordering attempts to sort senators from the most extreme to the most moderate and to the most extreme again. Figure 7 shows the Democrat senators and Figure 8 the Republicans.

We can observe that component 5 is the Democrat majority. It is the strongest overall component, yet quite uninfluential about the outcome. Component 4 are the moderate Democrats, and they seem distinctly more influential than the Democrats of the majority. Component 3 is a small group of Republican moderates. Component 2 is the Republican majority, the most influential bloc. Component 1 is the Republican minority, not very influential. Component 1 tends to be slightly more extreme than component 2 on the average, but the two components clearly cannot be unambiguously sorted.

9   Classification Experiments

DCA is not trying to capture those aspects of the text that are relevant for distinguishing one class of documents from another one. Assume our classification task is to distinguish newspaper articles on politics from all others. For DCA, modelling the distribution of a word such as 'the' is equally or more important than modelling the distribution of a highly pertinent word such as 'tsunami' in classifying news reports. Of course, this is only a single example of sub-optimality. Nevertheless, if there are several methods of reducing the dimensionality of text that all disregard the classification problem at hand, we can still compare them with respect to classification performance.

We used MPCA and tested its use in its role as a feature construction tool, a common use for PCA and ICA, and as a classification tool. For this, we used the 20 newsgroups collection described previously as well as the Reuters-21578 collection[7]. We employed the SVM^{light} V5.0 [Joa99] classifier with default settings. For classification, we added the class as a distinct multinomial (cf. Section 4.4) for the training data and left it empty for the test data, and then predicted the class value. Note that for performance and accuracy, SVM [Joa98] is a clear winner [LYRL04]. It is interesting to see how MPCA compares.

Each component can be seen as generating a number of words in each document. This number of component-generated words plays the same role in classification as does the number of lexemes in the document in ordinary classification. In both cases, we employed the `tf*idf` transformed word and component-generated word counts as feature values. Since SVM works with sparse data matrices, we assumed that a component is not present in a document if the number of words that a component would have generated is less than 0.01. The components alone do not yield a classification performance that would be competitive with SVM, as the label has no distinguished role in the fitting. However, we may add these component-words in the default bag of words, hoping that the conjunctions of words inherent to each component will help improve the classification performance.

For the Reuters collection, we used the ModApte split. For each of the 6 most frequent categories, we performed binary classification. Further results are disclosed in Table 2[8]. No major change was observed by adding 50 components to the original set of words. By performing classification on components alone, the results were inferior, even with a large number of components. In fact, with 300 components, the results were worse than with 200 components, probably because of over-fitting. Therefore, regardless of the number of components, the SVM performance with words cannot be reproduced by component-generated words in this collection. Classifying newsgroup articles into 20 categories proved more successful. We employed two replications of 5-fold cross validation, and we achieved the classification accuracy of 90.7% with 50 additional MPCA components, and 87.1% with SVM alone. Comparing the two confusion matrices, the most frequent mistakes caused by SVM+MPCA beyond those of SVM alone were predicting talk.politics.misc as sci.crypt (26 errors) and talk.religion.misc predicted as sci.electron (25 errors). On the other hand, the components helped better identify alt.atheism and talk.politics.misc, which were misclassified as talk.religion.misc (259 fewer errors) earlier. Also, talk.politics.misc and talk.religion.misc were not misclassified as talk.politics.gun (98 fewer errors). These 50 components were not very successful alone, resulting in 18.5% classification accuracy. By increasing the number of components to 100 and 300, the classification accuracy

---

[7] The Reuters-21578, Distribution 1.0 test collection is available from David D. Lewis' professional home page, currently: http://www.research.att.com/~lewis

[8] The numbers are percentages, and 'P/R' indicates precision/recall.

Table 5. SVM Classification Results

CAT	SVM ACC.	SVM P/R	SVM+MPCA ACC.	SVM+MPCA P/R
earn	98.58	98.5/97.1	98.45	98.2/97.1
acq	95.54	97.2/81.9	95.60	97.2/82.2
moneyfx	96.79	79.2/55.3	96.73	77.5/55.9
grain	98.94	94.5/81.2	98.70	95.7/74.5
crude	97.91	89.0/72.5	97.82	88.7/70.9
trade	98.24	79.2/68.1	98.36	81.0/69.8

CAT	MPCA (50 comp.) ACC.	MPCA (50 comp.) P/R	MPCA (200 comp.) ACC.	MPCA (200 comp.) P/R
earn	96.94	96.1/94.6	97.06	96.3/94.8
acq	92.63	93.6/71.1	92.33	95.3/68.2
moneyfx	95.48	67.0/33.0	96.61	76.0/54.7
grain	96.21	67.1/31.5	97.18	77.5/53.0
crude	96.57	81.1/52.4	96.79	86.1/52.4
trade	97.82	81.4/49.1	97.91	78.3/56.0

gradually increases to 25.0% and 34.3%. Therefore, many components are needed for general-purpose classification.

From these experiments, we can conclude that components may help with tightly coupled categories that require conjunctions of words (20 newsgroups), but not with the keyword-identifiable categories (Reuters). Judging from the ideas in [JB03], the components help in two cases: a) when the co-appearance of two words is more informative than the sum of informativeness of individual appearances of either word, and b) when the appearance of one word implies the appearance of another word, which does not always appear in the document.

10   Conclusion

In this article, we have presented a unifying framework for various approaches to discrete component analysis, presenting them as a model closely related to ICA but suited for sparse discrete data. We have shown the relationships between existing approaches here such as NMF, PLSI, LDA, MPCA and GaP. For instance, NMF with normalised results corresponds to an approximate maximum likelihood method for LDA, and GaP is the most general family of models. We have also presented the different algorithms available for three different cases, Gamma-Poisson, conditional Gamma-Poisson (allowing sparse component scores), and Dirichlet-multinomial. This extends a number of algorithms previous developed for MPCA and LDA to the general Gamma-Poisson model. Experiments with the MPCA software[9] show that a typical 3GHz desktop machine can build models in a few days with K in the hundreds for 3 gigabytes of text.

---

[9]   http://www.componentanalysis.org

These models share many similarities with both PCA and ICA, and are thus useful in a range of feature engineering tasks in machine learning and pattern recognition. A rich literature is alsoemerging extending the model in a variety of directions. This is as much caused by the surprising performance of the algorithms, as it is by the availability of general Gibbs sampling algorithms that allow sophisticated modelling.

Acknowledgments

Wray Buntine's work was supported by the Academy of Finland under the PROSE Project, by Finnish Technology Development Center (TEKES) under the Search-Ina-Box Project, by the IST Programme of the European Community under ALVIS Superpeer Semantic Search Engine (IST-1-002068-IP) and the PASCAL Network of Excellence (IST-2002-5006778). Aleks Jakulin was supported by the Slovenian Research Agency and by the IST Programme of the European Community under SEKT Semantically Enabled Knowledge Technologies (IST-1-506826-IP). The MPCA software used in the experiments was co-developed by a number of authors, reported at the code website. The experiments were supported by the document processing environment and test collections at the CoSCo group, and the information retrieval software YDIN of Sami Perttu. Both authors would also like to acknowledge very helpful comments from the referees.

References

[AGvR03] L. Azzopardi, M. Girolami, and K. van Risjbergen. Investigating the relationship between language model perplexity and ir precision-recall measures. In *SIGIR '03: Proceedings of the 26th annual international ACM SIGIR conference on Research and development in informaion retrieval*, pages 369–370, 2003.

[BJ04]   W. Buntine and A. Jakulin. Applying discrete PCA in data analysis. In *UAI-2004*, Banff, Canada, 2004.

[BKG03]   E. Bingham, A. Kabán, and M. Girolami. Topic identification in dynamical text by complexity pursuit. *Neural Process. Lett.*, 17(1):69–83, 2003.

[BNJ03]   D.M. Blei, A.Y. Ng, and M.I. Jordan. Latent Dirichlet allocation. *Journal of Machine Learning Research*, 3:993–1022, 2003.

[BPT04]   W.L. Buntine, S. Perttu, and V. Tuulos. Using discrete PCA on web pages. In *Workshop on Statistical Approaches to Web Mining, SAWM'04*, 2004. At ECML 2004.

[BS94]   J.M. Bernardo and A.F.M. Smith. *Bayesian Theory*. John Wiley, Chichester, 1994.

[Bun02]   W.L. Buntine. Variational extensions to EM and multinomial PCA. In *13th European Conference on Machine Learning (ECML'02)*, Helsinki, Finland, 2002.

[BYRN99] R. Baeza-Yates and B. Ribeiro-Neto. *Modern Information Retrieval*. Addison Wesley, 1999.

[Can04]   J. Canny. GaP: a factor model for discrete data. In *SIGIR 2004*, pages 122–129, 2004.

[CB90]   G. Casella and R.L. Berger. *Statistical Inference*. Wadsworth & Brooks/Cole, Belmont, CA, 1990.

[CB94]   B.S. Clarke and A.R. Barron. Jeffrey's prior is asymptotically least favorable under entropy risk. *Journal of Statistical Planning and Inference*, 41:37–60, 1994.

[CC95]   B.P. Carlin and S. Chib. Bayesian model choice via MCMC. *Journal of the Royal Statistical Society B*, 57:473–484, 1995.

[CDS01]   M. Collins, S. Dasgupta, and R.E. Schapire. A generalization of principal component analysis to the exponential family. In *NIPS*13*, 2001.

[CJR04]   J. D. Clinton, S. Jackman, and D. Rivers. The statistical analysis of roll call voting: A unified approach. *American Political Science Review*, 98(2):355–370, 2004.

[CR96]   G. Casella and C.P. Robert. Rao-Blackewellization of sampling schemes. *Biometrika*, 83(1):81–94, 1996.

[DDL+90]   S.C. Deerwester, S.T. Dumais, T.K. Landauer, G.W. Furnas, and R.A. Harshman. Indexing by latent semantic analysis. *Journal of the American Society of Information Science*, 41(6):391–407, 1990.

[deL03]   J. de Leeuw. Principal component analysis of binary data: Applications to roll-call-analysis. Technical Report 364, UCLA Department of Statistics, 2003.

[Dun94]   T. Dunning. Accurate methods for the statistics of surprise and coincidence. *Computational Linguistics*, 19(1):61–74, 1994.

[GB00]   Z. Ghahramani and M.J. Beal. Propagation algorithms for variational Bayesian learning. In *NIPS*, pages 507–513, 2000.

[GCSR95]   A. Gelman, J.B. Carlin, H.S. Stern, and D.B. Rubin. *Bayesian Data Analysis*. Chapman & Hall, 1995.

[GG05]   E. Gaussier and C. Goutte. Relation between PLSA and NMF and implications. In *SIGIR '05: Proceedings of the 28th annual international ACM SIGIR conference on Research and development in information retrieval*, pages 601–602. ACM Press, 2005.

[GS02]   T.L. Griffiths and M. Steyvers. A probabilistic approach to semantic representation. In *Proc. of the 24th Annual Conference of the Cognitive Science Society*, 2002.

[GS04]   T.L. Griffiths and M. Steyvers. Finding scientific topics. *PNAS Colloquium*, 2004.

[HB97]   T. Hofmann and J.M. Buhmann. Pairwise data clustering by deterministic annealing. *IEEE Transactions on Pattern Analysis and Machine Intelligence*, 19(1):1–14, 1997.

[HKO01]   A. Hyvärinen, J. Karhunen, and E. Oja. *Independent Component Analysis*. John Wiley & Sons, 2001.

[HLL83]   P. Holland, K. B. Laskey, and S. Leinhardt. Stochastic blockmodels: Some first steps. *Social Networks*, 5:109–137, 1983.

[HO00]   A. Hyvärinen and E. Oja. Independent component analysis: algorithms and applications. *Neural Netw.*, 13(4-5):411–430, 2000.

[Hof99]   T. Hofmann. Probabilistic latent semantic indexing. In *Research and Development in Information Retrieval*, pages 50–57, 1999.

[JB03]   A. Jakulin and I. Bratko. Analyzing attribute dependencies. In N. Lavrač, D. Gamberger, H. Blockeel, and L. Todorovski, editors, *PKDD 2003*, volume 2838 of *LNAI*, pages 229–240. Springer-Verlag, September 2003.

[Joa98]   T. Joachims. Text categorization with support vector machines: learning with many relevant features. In Claire Nédellec and Céline Rouveirol, editors, *Proceedings of ECML-98, 10th European Conference on Machine Learning*, number 1398, pages 137–142, Chemnitz, DE, 1998. Springer Verlag, Heidelberg, DE.

[Joa99]   T. Joachims. Making large-scale SVM learning practical. In B. Schölkopf, C. Burges, and A. Smola, editors, *Advances in Kernel Methods - Support Vector Learning*. MIT Press, 1999.

[LS99]   D. Lee and H. Seung. Learning the parts of objects by non-negative matrix factorization. *Nature*, 401:788–791, 1999.

[LYRL04]   D.D. Lewis, Y. Yand, T.G. Rose, and F. Li. Rcv1: A new benchmark collection for text categorization research. *Journal of Machine Learning Research*, 5:361–397, 2004.

[MKB79]   K.V. Mardia, J.T. Kent, and J.M. Bibby. *Multivariate Analysis*. Academic Press, 1979.

[ML02]   T. Minka and J. Lafferty. Expectation-propagation for the generative aspect model. In *UAI-2002*, Edmonton, 2002.

[MN89]   P. McCullagh and J.A. Nelder. *Generalized Linear Models*. Chapman and Hall, London, second edition, 1989.

[Poo00]   K.T. Poole. Non-parametric unfolding of binary choice data. *Political Analysis*, 8(3):211–232, 2000.

[PSD00]   J.K. Pritchard, M. Stephens, and P.J. Donnelly. Inference of population structure using multilocus genotype data. *Genetics*, 155:945–959, 2000.

[PTL93]   F. Pereira, N. Tishby, and L. Lee. Distributional clustering of English words. In *Proceedings of ACL-93*, June 1993.

[Ros89]   S.M. Ross. *Introduction to Probability Models*. Academic Press, fourth edition, 1989.

[Row98]   S. Roweis. EM algorithms for PCA and SPCA. In M.I. Jordan, M.J. Kearns, and S.A. Solla, editors, *Advances in Neural Information Processing Systems*, volume 10. The MIT Press, 1998.

[SN97]   T.A.B. Snijders and K. Nowicki. Estimation and prediction for stochastic block models for graphs with latent block structure. *Journal of Classification*, 14:75–100, 1997.

[TB99]   M.E. Tipping and C.M. Bishop. Probabilistic principal components analysis. *J. Roy. Statistical Society B*, 61(3):611–622, 1999.

[Tit]   D.M. Titterington. Some aspects of latent structure analysis. In this volume.

[vGv99]   P.G.M. van der Heijden, Z. Gilula, and L.A. van der Ark. An extended study into the relationship between correspondence analysis and latent class analysis. *Sociological Methodology*, 29:147–186, 1999.

[WM82]   M.A. Woodbury and K.G. Manton. A new procedure for analysis of medical classification. *Methods Inf Med*, 21:210–220, 1982.

[WMM05]   X. Wang, N. Mohanty, and A. McCallum. Group and topic discovery from relations and text. In *The 11th ACM SIGKDD International Conference on Knowledge Discovery and Data Mining Workshop on Link Discovery: Issues, Approaches and Applications (LinkKDD-05)*, pages 28–35, 2005.

[YYT05]   K. Yu, S. Yu, and V. Tresp. Dirichlet enhanced latent semantic analysis. In L.K. Saul, Y. Weiss, and L. Bottou, editors, *Proc. of the 10th International Workshop on Artificial Intelligence and Statistics*, 2005.

Overview and Recent Advances in Partial Least Squares

Roman Rosipal[1] and Nicole Krämer[2]

[1] Austrian Research Institute for Artificial Intelligence,
Freyung 6/6, A-1010 Vienna, Austria
roman.rosipal@ofai.at

[2] TU Berlin, Department of Computer Science and Electrical Engineering,
Franklinstraße 28/29, D-10587 Berlin, Germany
nkraemer@cs.tu-berlin.de

1 Introduction

Partial Least Squares (PLS) is a wide class of methods for modeling relations between sets of observed variables by means of latent variables. It comprises of regression and classification tasks as well as dimension reduction techniques and modeling tools. The underlying assumption of all PLS methods is that the observed data is generated by a system or process which is driven by a small number of latent (not directly observed or measured) variables. Projections of the observed data to its latent structure by means of PLS was developed by Herman Wold and coworkers [48, 49, 52].

PLS has received a great amount of attention in the field of chemometrics. The algorithm has become a standard tool for processing a wide spectrum of chemical data problems. The success of PLS in chemometrics resulted in a lot of applications in other scientific areas including bioinformatics, food research, medicine, pharmacology, social sciences, physiology–to name but a few [28, 25, 53, 29, 18, 22].

This chapter introduces the main concepts of PLS and provides an overview of its application to different data analysis problems. Our aim is to present a concise introduction, that is, a valuable guide for anyone who is concerned with data analysis.

In its general form PLS creates orthogonal score vectors (also called latent vectors or components) by maximising the covariance between different sets of variables. PLS dealing with two blocks of variables is considered in this chapter, although the PLS extensions to model relations among a higher number of sets exist [44, 46, 47, 48, 39]. PLS is similar to Canonical Correlation Analysis (CCA) where latent vectors with maximal correlation are extracted [24]. There are different PLS techniques to extract latent vectors, and each of them gives rise to a variant of PLS.

PLS can be naturally extended to regression problems. The predictor and predicted (response) variables are each considered as a block of variables. PLS then extracts the score vectors which serve as a new predictor representation

C. Saunders et al. (Eds.): SLSFS 2005, LNCS 3940, pp. 34–51, 2006.

and regresses the response variables on these new predictors. The natural asymmetry between predictor and response variables is reflected in the way in which score vectors are computed. This variant is known under the names of PLS1 (one response variable) and PLS2 (at least two response variables). PLS regression used to be overlooked by statisticians and is still considered rather an algorithm than a rigorous statistical model [14]. Yet within the last years, interest in the statistical properties of PLS has risen. PLS has been related to other regression methods like Principal Component Regression (PCR) [26] and Ridge Regression (RR) [16] and all these methods can be cast under a unifying approach called continuum regression [40, 9]. The effectiveness of PLS has been studied theoretically in terms of its variance [32] and its shrinkage properties [12, 21, 7]. The performance of PLS is investigated in several simulation studies [11, 1].

PLS can also be applied to classification problems by encoding the class membership in an appropriate indicator matrix. There is a close connection of PLS for classification to Fisher Discriminant Analysis (FDA) [4]. PLS can be applied as a discrimination tool and dimension reduction method–similar to Principal Component Analysis (PCA). After relevant latent vectors are extracted, an appropriate classifier can be applied. The combination of PLS with Support Vector Machines (SVM) has been studied in [35].

Finally, the powerful machinery of kernel-based learning can be applied to PLS. Kernel methods are an elegant way of extending linear data analysis tools to nonlinear problems [38].

2 Partial Least Squares

Consider the general setting of a linear PLS algorithm to model the relation between two data sets (blocks of variables). Denote by $\mathcal{X} \subset \mathcal{R}^N$ an N-dimensional space of variables representing the first block and similarly by $\mathcal{Y} \subset \mathcal{R}^M$ a space representing the second block of variables. PLS models the relations between these two blocks by means of score vectors. After observing n data samples from each block of variables, PLS decomposes the $(n \times N)$ matrix of zero-mean variables $\mathbf{X}$ and the $(n \times M)$ matrix of zero-mean variables $\mathbf{Y}$ into the form

$$\begin{aligned} \mathbf{X} &= \mathbf{T}\mathbf{P}^T + \mathbf{E} \\ \mathbf{Y} &= \mathbf{U}\mathbf{Q}^T + \mathbf{F} \end{aligned} \qquad (1)$$

where the $\mathbf{T}$, $\mathbf{U}$ are $(n \times p)$ matrices of the p extracted score vectors (components, latent vectors), the $(N \times p)$ matrix $\mathbf{P}$ and the $(M \times p)$ matrix $\mathbf{Q}$ represent matrices of loadings and the $(n \times N)$ matrix $\mathbf{E}$ and the $(n \times M)$ matrix $\mathbf{F}$ are the matrices of residuals. The PLS method, which in its classical form is based on the nonlinear iterative partial least squares (NIPALS) algorithm [47], finds weight vectors $\mathbf{w}, \mathbf{c}$ such that

$$[cov(\mathbf{t}, \mathbf{u})]^2 = [cov(\mathbf{X}\mathbf{w}, \mathbf{Y}\mathbf{c})]^2 = max_{|\mathbf{r}|=|\mathbf{s}|=1}[cov(\mathbf{X}\mathbf{r}, \mathbf{Y}\mathbf{s})]^2 \qquad (2)$$

where $cov(\mathbf{t}, \mathbf{u}) = \mathbf{t}^T\mathbf{u}/n$ denotes the sample covariance between the score vectors $\mathbf{t}$ and $\mathbf{u}$. The NIPALS algorithm starts with random initialisation of the

$\mathcal{Y}$-space score vector $\mathbf{u}$ and repeats a sequence of the following steps until convergence.

1) $\mathbf{w} = \mathbf{X}^T\mathbf{u}/(\mathbf{u}^T\mathbf{u})$ 4) $\mathbf{c} = \mathbf{Y}^T\mathbf{t}/(\mathbf{t}^T\mathbf{t})$
2) $\|\mathbf{w}\| \rightarrow 1$ 5) $\|\mathbf{c}\| \rightarrow 1$
3) $\mathbf{t} = \mathbf{Xw}$ 6) $\mathbf{u} = \mathbf{Yc}$

Note that $\mathbf{u} = \mathbf{y}$ if $M = 1$, that is, $\mathbf{Y}$ is a one-dimensional vector that we denote by $\mathbf{y}$. In this case the NIPALS procedure converges in a single iteration.

It can be shown that the weight vector $\mathbf{w}$ also corresponds to the first eigenvector of the following eigenvalue problem [17]

$$\mathbf{X}^T\mathbf{Y}\mathbf{Y}^T\mathbf{Xw} = \lambda\mathbf{w} \tag{3}$$

The $\mathcal{X}$- and $\mathcal{Y}$-space score vectors $\mathbf{t}$ and $\mathbf{u}$ are then given as

$$\mathbf{t} = \mathbf{Xw} \quad \text{and} \quad \mathbf{u} = \mathbf{Yc} \tag{4}$$

where the weight vector $\mathbf{c}$ is define in steps 4 and 5 of NIPALS. Similarly, eigenvalue problems for the extraction of $\mathbf{t}$, $\mathbf{u}$ or $\mathbf{c}$ estimates can be derived [17]. The user then solves for one of these eigenvalue problems and the other score or weight vectors are readily computable using the relations defined in NIPALS.

2.1   Forms of PLS

PLS is an iterative process. After the extraction of the score vectors $\mathbf{t}$, $\mathbf{u}$ the matrices $\mathbf{X}$ and $\mathbf{Y}$ are deflated by subtracting their rank-one approximations based on $\mathbf{t}$ and $\mathbf{u}$. Different forms of deflation define several variants of PLS.

Using equations (1) the vectors of loadings $\mathbf{p}$ and $\mathbf{q}$ are computed as coefficients of regressing $\mathbf{X}$ on $\mathbf{t}$ and $\mathbf{Y}$ on $\mathbf{u}$, respectively

$$\mathbf{p} = \mathbf{X}^T\mathbf{t}/(\mathbf{t}^T\mathbf{t}) \quad \text{and} \quad \mathbf{q} = \mathbf{Y}^T\mathbf{u}/(\mathbf{u}^T\mathbf{u})$$

PLS Mode A: The PLS Mode A is based on rank-one deflation of individual block matrices using the corresponding score and loading vectors. In each iteration of PLS Mode A the $\mathbf{X}$ and $\mathbf{Y}$ matrices are deflated

$$\mathbf{X} = \mathbf{X} - \mathbf{tp}^T \quad \text{and} \quad \mathbf{Y} = \mathbf{Y} - \mathbf{uq}^T$$

This approach was originally designed by Herman Wold [47] to model the relations between the different sets (blocks) of data. In contrast to the PLS regression approach, discussed next, the relation between the two blocks is symmetric. As such this approach seems to be appropriate for modeling existing relations between sets of variables in contrast to prediction purposes. In this way PLS Mode A is similar to CCA. Wegelin [45] discusses and compares properties of both methods. The connection between PLS and CCA from the point of an optimisation criterion involved in each method is discussed in Section 2.2.

PLS1, PLS2: PLS1 (one of the block of data consists of a single variable) and PLS2 (both blocks are multidimensional) are used as PLS regression methods. These variants of PLS are the most frequently used PLS approaches. The relationship between $\mathbf{X}$ and $\mathbf{Y}$ is asymmetric. Two assumptions are made: i) the score vectors $\{\mathbf{t}_i\}_{i=1}^{p}$ are good predictors of $\mathbf{Y}$; p denotes the number of extracted score vectors–PLS iterations ii) a linear inner relation between the scores vectors $\mathbf{t}$ and $\mathbf{u}$ exists; that is,

$$\mathbf{U} = \mathbf{TD} + \mathbf{H} \tag{5}$$

where $\mathbf{D}$ is the $(p \times p)$ diagonal matrix and $\mathbf{H}$ denotes the matrix of residuals. The asymmetric assumption of the predictor–predicted variable(s) relation is transformed into a deflation scheme where the predictor space, say $\mathbf{X}$, score vectors $\{\mathbf{t}_i\}_{i=1}^{p}$ are good predictors of $\mathbf{Y}$. The score vectors are then used to deflate $\mathbf{Y}$, that is, a component of the regression of $\mathbf{Y}$ on $\mathbf{t}$ is removed from $\mathbf{Y}$ at each iteration of PLS

$$\mathbf{X} = \mathbf{X} - \mathbf{t}\mathbf{p}^T \quad \text{and} \quad \mathbf{Y} = \mathbf{Y} - \mathbf{t}\mathbf{t}^T\mathbf{Y}/(\mathbf{t}^T\mathbf{t}) = \mathbf{Y} - \mathbf{t}\mathbf{c}^T$$

where we consider not scaled to unit norm weight vectors $\mathbf{c}$ defined in step 4 of NIPALS. This deflation scheme guarantees mutual orthogonality of the extracted score vectors $\{\mathbf{t}_i\}_{i=1}^{p}$ [17]. Note that in PLS1 the deflation of $\mathbf{y}$ is technically not needed during the iterations of PLS [17].

Singular values of the cross-product matrix $\mathbf{X}^T\mathbf{Y}$ correspond to the sample covariance values [17]. Then the deflation scheme of extracting one component at a time has also the following interesting property. The first singular value of the deflated cross-product matrix $\mathbf{X}^T\mathbf{Y}$ at iteration $i + 1$ is greater or equal than the second singular value of $\mathbf{X}^T\mathbf{Y}$ at iteration i [17]. This result can be also applied to the relation of eigenvalues of (3) due to the fact that (3) corresponds to the singular value decomposition of the transposed cross-product matrix $\mathbf{X}^T\mathbf{Y}$. In particular, the PLS1 and PLS2 algorithms differ from the computation of all eigenvectors of (3) in one step.

PLS-SB: As outlined at the end of the previous paragraph the computation of all eigenvectors of (3) at once would define another form of PLS. This computation involves a sequence of implicit rank-one deflations of the overall cross-product matrix. This form of PLS was used in [36] and in accordance with [45] it is denoted as PLS-SB. In contrast to PLS1 and PLS2, the extracted score vectors $\{\mathbf{t}_i\}_{i=1}^{p}$ are in general not mutually orthogonal.

SIMPLS: To avoid deflation steps at each iteration of PLS1 and PLS2, de Jong [8] has introduced another form of PLS denoted SIMPLS. The SIMPLS approach directly finds the weight vectors $\{\tilde{\mathbf{w}}\}_{i=1}^{p}$ which are applied to the original not deflated matrix $\mathbf{X}$. The criterion of the mutually orthogonal score vectors $\{\tilde{\mathbf{t}}\}_{i=1}^{p}$ is kept. It has been shown that SIMPLS is equal to PLS1 but differs from PLS2 when applied to the multidimensional matrix $\mathbf{Y}$ [8].

2.2   PCA, CCA and PLS

There exists a variety of different projection methods to latent variables. Among others widely used, PCA and CCA belong to this category. The connections between PCA, CCA and PLS can be seen through the optimisation criterion they use to define projection directions. PCA projects the original variables onto a direction of maximal variance called principal direction. Following the notation of (2), the optimisation criterion of PCA can be written as

$$max_{|\mathbf{r}|=1}[var(\mathbf{Xr})]$$

where $var(\mathbf{t}) = \mathbf{t}^T\mathbf{t}/n$ denotes the sample variance. Similarly CCA finds the direction of maximal correlation solving the following optimisation problem

$$max_{|\mathbf{r}|=|\mathbf{s}|=1}[corr(\mathbf{Xr}, \mathbf{Ys})]^2$$

where $[corr(\mathbf{t}, \mathbf{u})]^2 = [cov(\mathbf{t}, \mathbf{u})]^2/var(\mathbf{t})var(\mathbf{u})$ denotes the sample squared correlation. It is easy to see that the PLS criterion (2)

$$max_{|\mathbf{r}|=|\mathbf{s}|=1}[cov(\mathbf{Xr}, \mathbf{Ys})]^2 = max_{|\mathbf{r}|=|\mathbf{s}|=1}var(\mathbf{Xr})[corr(\mathbf{Xr}, \mathbf{Ys})]^2var(\mathbf{Ys}) \quad (6)$$

represents a form of CCA where the criterion of maximal correlation is balanced with the requirement to explain as much variance as possible in both $\mathcal{X}$- and $\mathcal{Y}$-spaces. Note that in the case of a one-dimensional $\mathcal{Y}$-space only the $\mathcal{X}$-space variance is involved.

The relation between CCA and PLS can be also seen through the concept of canonical ridge analysis introduced in [41]. Consider the following optimisation problem

$$\max_{|\mathbf{r}|=|\mathbf{s}|=1} \frac{cov(\mathbf{Xr}, \mathbf{Ys})^2}{([1 - \gamma_{\mathbf{X}}] var(\mathbf{Xr}) + \gamma_{\mathbf{X}}) ([1 - \gamma_{\mathbf{Y}}] var(\mathbf{Ys}) + \gamma_{\mathbf{Y}})}$$

with $0 \leq \gamma_{\mathbf{X}}, \gamma_{\mathbf{Y}} \leq 1$ representing regularisation terms. The corresponding eigenvalue problem providing the solution to this optimisation criterion is given as

$$([1 - \gamma_{\mathbf{X}}]\mathbf{X}^T\mathbf{X} + \gamma_{\mathbf{X}}\mathbf{I})^{-1}\mathbf{X}^T\mathbf{Y}([1 - \gamma_{\mathbf{Y}}]\mathbf{Y}^T\mathbf{Y} + \gamma_{\mathbf{Y}}\mathbf{I})^{-1}\mathbf{Y}^T\mathbf{Xw} = \lambda\mathbf{w} \quad (7)$$

where $\mathbf{w}$ represents a weight vector for the projection of the original $\mathcal{X}$-space data into a latent space.[1] There are two cornerstone solutions of this eigenvalue problem: i) for $\gamma_{\mathbf{X}} = 0, \gamma_{\mathbf{Y}} = 0$ the solution of CCA is obtained [24] ii) for $\gamma_{\mathbf{X}} = 1, \gamma_{\mathbf{Y}} = 1$ the PLS eigenvalue problem (3) is recovered. By continuous changing of $\gamma_{\mathbf{X}}, \gamma_{\mathbf{Y}}$ solutions lying between these two cornerstones are obtained. In Figure 1 the $\mathbf{w}$ directions for two-class problem as found by PLS, CCA and regularised CCA ($\gamma_{\mathbf{X}} = 0.99, \gamma_{\mathbf{Y}} = 0$) are plotted.

Another interesting setting is $\gamma_{\mathbf{X}} = 1, \gamma_{\mathbf{Y}} = 0$ which represents a form of orthonormalised PLS where the $\mathcal{Y}$-space data variance does not influence the

---

[1] In the analogous way the eigenvalue problem for the projections of the $\mathcal{Y}$-space data can be formulated.

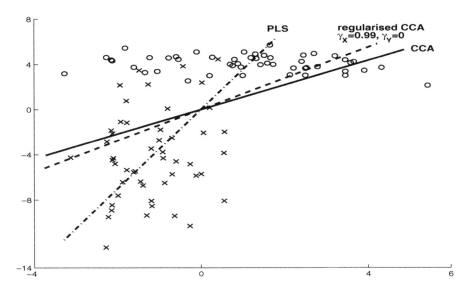

Fig. 1. An example of the weight vector $\mathbf{w}$ directions as found by CCA (solid line), PLS (dash-dotted line) and regularised CCA (dashed line) given by (7) with $\gamma_{\mathbf{X}} = 0.99$ and $\gamma_{\mathbf{Y}} = 0$. Circle and cross samples represent two Gaussian distributed classes with different sample means and covariances.

final PLS solution (similarly the $\mathcal{X}$-space variance can be ignored by setting $\gamma_{\mathbf{X}} = 0, \gamma_{\mathbf{Y}} = 1$) [53]. Note that in the case of one-dimensional $\mathbf{Y}$ matrix and for $\gamma_{\mathbf{X}} \in (0, 1)$ the ridge regression solution is obtained [41, 16]. Finally let us stress that, in general, CCA is solved in a way similar to PLS-SB, that is, eigenvectors and eigenvalues of (7) are extracted at once by an implicit deflation of the cross-product matrix $\mathbf{X}^T\mathbf{Y}$. This is in contrast to the PLS1 and PLS2 approaches where different deflation scheme is considered.

3   PLS Regression

As mentioned in the previous section, PLS1 and PLS2 can be used to solve linear regression problems. Combining assumption (5) of a linear relation between the scores vectors $\mathbf{t}$ and $\mathbf{u}$ with the decomposition of the $\mathbf{Y}$ matrix, equation (1) can be written as

$$\mathbf{Y} = \mathbf{TDQ}^T + (\mathbf{HQ}^T + \mathbf{F})$$

This defines the equation

$$\mathbf{Y} = \mathbf{TC}^T + \mathbf{F}^* \tag{8}$$

where $\mathbf{C}^T = \mathbf{DQ}^T$ now denotes the $(p \times M)$ matrix of regression coefficients and $\mathbf{F}^* = \mathbf{HQ}^T + \mathbf{F}$ is the residual matrix. Equation (8) is simply the decomposition of $\mathbf{Y}$ using ordinary least squares regression with orthogonal predictors $\mathbf{T}$.

We now consider orthonormalised score vectors $\mathbf{t}$, that is, $\mathbf{T}^T\mathbf{T} = \mathbf{I}$, and the matrix $\mathbf{C} = \mathbf{Y}^T\mathbf{T}$ of the not scaled to length one weight vectors $\mathbf{c}$. It is useful to redefine equation (8) in terms of the original predictors $\mathbf{X}$. To do this, we use the relationship [23]

$$\mathbf{T} = \mathbf{X}\mathbf{W}(\mathbf{P}^T\mathbf{W})^{-1}$$

where $\mathbf{P}$ is the matrix of loading vectors defined in (1). Plugging this relation into (8), we yield

$$\mathbf{Y} = \mathbf{X}\mathbf{B} + \mathbf{F}^*$$

where $\mathbf{B}$ represents the matrix of regression coefficients

$$\mathbf{B} = \mathbf{W}(\mathbf{P}^T\mathbf{W})^{-1}\mathbf{C}^T = \mathbf{X}^T\mathbf{U}(\mathbf{T}^T\mathbf{X}\mathbf{X}^T\mathbf{U})^{-1}\mathbf{T}^T\mathbf{Y}$$

For the last equality, the relations among $\mathbf{T}$, $\mathbf{U}$, $\mathbf{W}$ and $\mathbf{P}$ are used [23, 17, 33]. Note that different scalings of the individual score vectors $\mathbf{t}$ and $\mathbf{u}$ do not influence the $\mathbf{B}$ matrix. For training data the estimate of PLS regression is

$$\hat{\mathbf{Y}} = \mathbf{X}\mathbf{B} = \mathbf{T}\mathbf{T}^T\mathbf{Y} = \mathbf{T}\mathbf{C}^T$$

and for testing data we have

$$\hat{\mathbf{Y}}_t = \mathbf{X}_t\mathbf{B} = \mathbf{T}_t\mathbf{T}^T\mathbf{Y} = \mathbf{T}_t\mathbf{C}^T$$

where $\mathbf{X}_t$ and $\mathbf{T}_t = \mathbf{X}_t\mathbf{X}^T\mathbf{U}(\mathbf{T}^T\mathbf{X}\mathbf{X}^T\mathbf{U})^{-1}$ represent the matrices of testing data and score vectors, respectively.

3.1   Algebraic Interpretation of Linear Regression

In this paragraph, we only consider PLS1, that is, the output data $\mathbf{y}$ is a one-dimensional vector. The linear regression model is usually subsumed in the relation

$$\mathbf{y} = \mathbf{X}\mathbf{b} + \mathbf{e} \tag{9}$$

with $\mathbf{b}$ the unknown regression vector and $\mathbf{e}$ a vector of independent identically distributed noise with $var(\mathbf{e}) = \sigma^2$. In what follows, we will make intensive use of the singular value decomposition of $\mathbf{X}$

$$\mathbf{X} = \mathbf{V}\mathbf{\Sigma}\mathbf{S}^T \tag{10}$$

with $\mathbf{V}$ and $\mathbf{S}$ orthonormal matrices and $\mathbf{\Sigma}$ a diagonal matrix that consists of the singular values of $\mathbf{X}$. The matrix $\mathbf{\Lambda} = \mathbf{\Sigma}^2$ is diagonal with elements λ_i. Set

$$\mathbf{A} \equiv \mathbf{X}^T\mathbf{X} = \mathbf{S}\mathbf{\Lambda}\mathbf{S}^T \quad \text{and} \quad \mathbf{z} \equiv \mathbf{X}^T\mathbf{y}$$

The ordinary least squares (OLS) estimator $\hat{\mathbf{b}}_{OLS}$ is the solution of

$$arg \min_{\mathbf{b}} \|\mathbf{y} - \mathbf{X}\mathbf{b}\|^2$$

This problem is equivalent to computing the solution of the normal equations

$$\mathbf{A}\mathbf{b} = \mathbf{z} \tag{11}$$

Using the pseudoinverse of $\mathbf{A}^-$, it follows (recall (10)) that

$$\hat{\mathbf{b}}_{OLS} = \mathbf{A}^- \mathbf{z} = \sum_{i=1}^{rk(\mathbf{A})} \frac{\mathbf{v}_i^T \mathbf{y}}{\sqrt{\lambda_i}} \mathbf{s}_i = \sum_{i=1}^{rk(\mathbf{A})} \hat{\mathbf{b}}_i$$

where

$$\hat{\mathbf{b}}_i = \frac{\mathbf{v}_i^T \mathbf{y}}{\sqrt{\lambda_i}} \mathbf{s}_i$$

is the component of $\hat{\mathbf{b}}_{OLS}$ along $\mathbf{v}_i$ and $rk(.)$ denotes the rank of a matrix.

A lot of linear regression estimators are approximate solutions of the equation (11). The PCR estimator that regresses $\mathbf{y}$ on the first p principal components $\mathbf{v}_1, \ldots, \mathbf{v}_p$ is

$$\hat{\mathbf{b}}_{PCR} = \sum_{i=1}^{p} \hat{\mathbf{b}}_i$$

The RR estimator [41, 16] is of the form

$$\hat{\mathbf{b}}_{RR} = (\mathbf{A} + \gamma \mathbf{I})^{-1} \mathbf{z} = \sum_{i=1}^{rk(\mathbf{A})} \frac{\lambda_i}{\lambda_i + \gamma} \hat{\mathbf{b}}_i$$

with $\gamma > 0$ the ridge parameter.

It can be shown that the PLS algorithm is equivalent to the conjugate gradient method [15]. This is a procedure that iteratively computes approximate solutions of (11) by minimising the quadratic function

$$\frac{1}{2} \mathbf{b}^T \mathbf{A} \mathbf{b} - \mathbf{z}^T \mathbf{b}$$

along directions that are $\mathbf{A}$-orthogonal. The approximate solution obtained after p steps is equal to the PLS estimator obtained after p iterations.

The conjugate gradient algorithm is in turn closely related to the Lanczos algorithm [19], a method for approximating eigenvalues. The space spanned by the columns of

$$\mathbf{K} = (\mathbf{z}, \mathbf{A}\mathbf{z}, \ldots, \mathbf{A}^{p-1}\mathbf{z})$$

is called the p-dimensional Krylov space of $\mathbf{A}$ and $\mathbf{z}$. We denote this Krylov space by $\mathcal{K}$. In the Lanczos algorithm, an orthogonal basis

$$\mathbf{W} = (\mathbf{w}_1, \ldots, \mathbf{w}_p) \tag{12}$$

of $\mathcal{K}$ is computed. The linear map $\mathbf{A}$ restricted to $\mathcal{K}$ for an element $\mathbf{k} \in \mathcal{K}$ is defined as the orthogonal projection of $\mathbf{A}\mathbf{k}$ onto the space $\mathcal{K}$. The map is represented by the $p \times p$ matrix

$$\mathbf{L} = \mathbf{W}^T \mathbf{A} \mathbf{W}$$

This matrix is tridiagonal. Its p eigenvector-eigenvalue pairs

$$(\mathbf{r}_i, \mu_i) \tag{13}$$

are called Ritz pairs. They are the best approximation of the eigenpairs of $\mathbf{A}$ given only the information that is encoded in $\mathcal{K}$ [30].

The weight vectors $\mathbf{w}$ in (2) of PLS1 are identical to the basis vectors in (12). In particular, the weight vectors are a basis of the Krylov space and the PLS estimator is the solution of the optimisation problem

$$arg \min_{\mathbf{b}} \|\mathbf{y} - \mathbf{Xb}\|^2$$

$$subject\ to \quad \mathbf{b} \in \mathcal{K}$$

In this sense, PLS1 can be viewed as a regularised least squares fit.

A good references for the Lanczos method and the conjugate gradient method is [30]. The connection to PLS is well-elaborated in [31].

3.2   Shrinkage Properties of PLS Regression

One possibility to evaluate the quality of an estimator $\hat{\mathbf{b}}$ for $\mathbf{b}$ is to determine its Mean Squared Error (MSE), which is defined as

$$\mathrm{MSE}(\hat{\mathbf{b}}) = E\left[\left(\hat{\mathbf{b}} - \mathbf{b}\right)^T \left(\hat{\mathbf{b}} - \mathbf{b}\right)\right]$$

$$= \left(E\left[\hat{\mathbf{b}}\right] - \mathbf{b}\right)^T \left(E\left[\hat{\mathbf{b}}\right] - \mathbf{b}\right) + E\left[\left(\hat{\mathbf{b}} - E\left[\hat{\mathbf{b}}\right]\right)^T \left(\hat{\mathbf{b}} - E\left[\hat{\mathbf{b}}\right]\right)\right]$$

This is the well-known bias-variance decomposition of MSE. The first part is the squared bias and the second part is the variance term.

It is well known that the OLS estimator has no bias (if $\mathbf{b} \in range(\mathbf{A})$). The variance term depends on the non-zero eigenvalues of $\mathbf{A}$: if some eigenvalues are very small, the variance of $\hat{\mathbf{b}}_{OLS}$ can be very high, which leads to a high MSE value. Note that small eigenvalues λ_i of $\mathbf{A}$ correspond to principal directions $\mathbf{v}_i$ of $\mathbf{X}$ that have a low sample spread.

One possibility to decrease MSE is to modify the OLS estimator by shrinking the directions of the OLS estimator that are responsible for a high variance. In general, a shrinkage estimator for $\mathbf{b}$ is of the form

$$\hat{\mathbf{b}}_{shr} = \sum_{i=1}^{rk(\mathbf{A})} f(\lambda_i)\hat{\mathbf{b}}_i \qquad (14)$$

where $f(.)$ is some real-valued function. The values $f(\lambda_i)$ are called shrinkage factors. Examples are PCR

$$f(\lambda_i) = \begin{cases} 1\ , & i \leq p \\ 0\ , & i > p \end{cases}$$

and RR

$$f(\lambda_i) = \frac{\lambda_i}{\lambda_i + \gamma}$$

If the factors in (14) do not depend on $\mathbf{y}$, that is, $\hat{\mathbf{b}}_{shr}$ is linear in $\mathbf{y}$, any factor $f(\lambda_i) \neq 1$ increases the bias of the i-th component. The variance of the i-th component decreases for $|f(\lambda_i)| < 1$ and increases for $|f(\lambda_i)| > 1$. The OLS estimator is shrunk in the hope that the increase in bias is small compared to the decrease in variance.

The PLS estimator is a shrinkage estimator as well. Its shrinkage factors are closely related to the Ritz pairs (13). The shrinkage factors $f(\lambda_i)$ that correspond to the estimator $\hat{\mathbf{b}}_{PLS}$ after p iterations of PLS are [21, 31]

$$f(\lambda_i) = 1 - \prod_{j=1}^{p} \left(1 - \frac{\lambda_i}{\mu_j}\right)$$

The shrinkage factors have some remarkable properties [7, 21]. Most importantly, $f(\lambda_i) > 1$ can occur for certain combinations of i and p. Note however that the PLS estimator is not linear in $\mathbf{y}$. The factors $f(\lambda_i)$ depend on the eigenvalues (13) of the matrix $\mathbf{L}$ and $\mathbf{L}$ in turn depends, via $\mathbf{z}$, on $\mathbf{y}$. It is therefore not clear in which way this shrinkage behaviour influences MSE of PLS1.

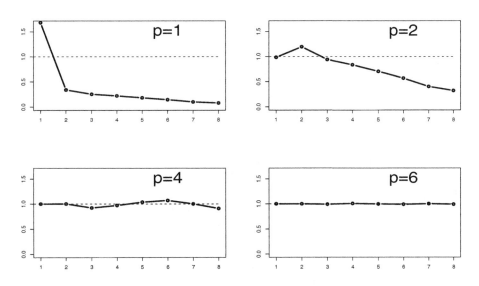

Fig. 2. An illustration of the shrinkage behaviour of PLS1. The $\mathbf{X}$ matrix contains eight variables. The eigenvalues of $\mathbf{A} \equiv \mathbf{X}^T\mathbf{X}$ are enumerated in decreasing order, $\lambda_1 \geq \lambda_2 \geq \ldots$ and the shrinkage factors $f(\lambda_i)$ are plotted as a function of i. The amount of absolute shrinkage $|1 - f(\lambda_i)|$ is particularly prominent if p is small.

4   PLS Discrimination and Classification

PLS has been used for discrimination and classification purposes. The close connection between FDA, CCA and PLS in the discrimination and classification scenario is described in this section.

Consider a set of n samples $\{\mathbf{x}_i \in \mathcal{X} \subset \mathcal{R}^N\}_{i=1}^n$ representing the data from g classes (groups). Now define the $(n \times g - 1)$ class membership matrix $\mathbf{Y}$ to be

$$
\mathbf{Y} = \begin{pmatrix}
\mathbf{1}_{n_1} & \mathbf{0}_{n_1} & \cdots & \mathbf{0}_{n_1} \\
\mathbf{0}_{n_2} & \mathbf{1}_{n_2} & \cdots & \mathbf{0}_{n_2} \\
\vdots & \vdots & \ddots & \mathbf{1}_{n_{g-1}} \\
\mathbf{0}_{n_g} & \mathbf{0}_{n_g} & \cdots & \mathbf{0}_{n_g}
\end{pmatrix}
$$

where $\{n_i\}_{i=1}^g$ denotes the number of samples in each class, $\sum_{i=1}^g n_i = n$ and $\mathbf{0}_{n_i}$ and $\mathbf{1}_{n_i}$ are $(n_i \times 1)$ vectors of all zeros and ones, respectively. Let

$$
\mathbf{S_X} = \frac{1}{n-1}\mathbf{X}^T\mathbf{X} \quad , \quad \mathbf{S_Y} = \frac{1}{n-1}\mathbf{Y}^T\mathbf{Y} \quad \text{and} \quad \mathbf{S_{XY}} = \frac{1}{n-1}\mathbf{X}^T\mathbf{Y}
$$

be the sample estimates of the covariance matrices $\mathbf{\Sigma_X}$ and $\mathbf{\Sigma_Y}$, respectively, and the cross-product covariance matrix $\mathbf{\Sigma_{XY}}$. Again, the matrices $\mathbf{X}$ and $\mathbf{Y}$ are considered to be zero-mean. Furthermore, let

$$
\mathbf{H} = \sum_{i=1}^g n_i(\bar{\mathbf{x}}_i - \bar{\mathbf{x}})(\bar{\mathbf{x}}_i - \bar{\mathbf{x}})^T \quad , \quad \mathbf{E} = \sum_{i=1}^g \sum_{j=1}^{n_i}(\mathbf{x}_i^j - \bar{\mathbf{x}}_i)(\mathbf{x}_i^j - \bar{\mathbf{x}}_i)^T
$$

represent the *among-classes* and *within-classes* sums-of-squares, where $\mathbf{x}_i^j$ represents an N-dimensional vector for the j-th sample in the i-th class and

$$
\bar{\mathbf{x}}_i = \frac{1}{n_i}\sum_{j=1}^{n_i}\mathbf{x}_i^j \quad \text{and} \quad \bar{\mathbf{x}} = \frac{1}{n}\sum_{i=1}^g \sum_{j=1}^{n_i}\mathbf{x}_i^j
$$

Fisher developed a discrimination method based on a linear projection of the input data such that among-classes variance is maximised relative to the within-classes variance. The directions onto which the input data are projected are given by the eigenvectors $\mathbf{a}$ of the eigenvalue problem

$$
\mathbf{E}^{-1}\mathbf{H}\mathbf{a} = \lambda\mathbf{a}
$$

In the case of discriminating multi-normally distributed classes with the same covariance matrices, FDA finds the same discrimination directions as linear discriminant analysis using Bayes theorem to estimate posterior class probabilities. This is the method that provides the discrimination rule with minimal expected misclassification error [24, 13].

The fact that the Fisher's discrimination directions are identical to the directions given by CCA using a dummy matrix $\mathbf{Y}$ for group membership was first recognised in [5]. The connections between PLS and CCA have been methodically studied in [4]. Among other, the authors argue that the $\mathcal{Y}$-space penalty $var(\mathbf{Ys})$ is not meaningful and suggested to remove it from (6) in the PLS discrimination scenario. As mentioned in Section 2.2 this modification leads to a special case of the previously proposed orthonormalised PLS method [53] using the indicator matrix $\mathbf{Y}$. The eigenvalue problem (3) in the case of orthonormalised PLS is transformed into

$$\mathbf{X}^T\mathbf{Y}(\mathbf{Y}^T\mathbf{Y})^{-1}\mathbf{Y}^T\mathbf{X}\mathbf{w} = \mathbf{X}^T\tilde{\mathbf{Y}}\tilde{\mathbf{Y}}^T\mathbf{X}\mathbf{w} = \lambda\mathbf{w} \qquad (15)$$

where

$$\tilde{\mathbf{Y}} = \mathbf{Y}(\mathbf{Y}^T\mathbf{Y})^{-1/2}$$

represents a matrix of uncorrelated and normalised output variables. Using the following relation [4, 35]

$$(n-1)\mathbf{S}_{\mathbf{XY}}\mathbf{S}_{\mathbf{Y}}^{-1}\mathbf{S}_{\mathbf{XY}}^T = \mathbf{H}$$

the eigenvectors of (15) are equivalent to the eigensolutions of

$$\mathbf{H}\mathbf{w} = \lambda\mathbf{w} \qquad (16)$$

Thus, this modified PLS method is based on eigensolutions of the among-classes sum-of-squares matrix $\mathbf{H}$ which connects this approach to CCA or equivalently to FDA.

Interestingly, in the case of two-class discrimination the direction of the first orthonormalised PLS score vector $\mathbf{t}$ is identical with the first score vector found by either the PLS1 or PLS-SB methods. This immediately follows from the fact that $\mathbf{Y}^T\mathbf{Y}$ is a number in this case. In this two-class scenario $\mathbf{X}^T\mathbf{Y}$ is of a rank-one matrix and PLS-SB extracts only one score vector $\mathbf{t}$. In contrast, orthonormalised PLS can extract additional score vectors, up to the rank of $\mathbf{X}$, each being similar to directions computed with CCA or FDA on deflated feature space matrices. Thus, PLS provide more principled dimensionality reduction in comparison to PCA based on the criterion of maximum data variation in the $\mathcal{X}$-space alone.

In the case of multi-class discrimination the rank of the $\mathbf{Y}$ matrix is equal to $g-1$ which determines the maximum number of score vectors that may be extracted by the orthonormalised PLS-SB method.[2] Again, similar to the one-dimensional output scenario the deflation of the $\mathbf{Y}$ matrix at each step can be done using the score vectors $\mathbf{t}$ of PLS2. Consider this deflation scheme in the $\mathcal{X}$- and $\mathcal{Y}$-spaces

$$\mathbf{X}_d = \mathbf{X} - \mathbf{t}\mathbf{p}^T = (\mathbf{I} - \mathbf{t}\mathbf{t}^T/(\mathbf{t}^T\mathbf{t}))\mathbf{X} = \mathbf{P}_d\mathbf{X}$$
$$\tilde{\mathbf{Y}}_d = \mathbf{P}_d\tilde{\mathbf{Y}}$$

where $\mathbf{P}_d = \mathbf{P}_d^T\mathbf{P}_d$ represents a projection matrix. Using these deflated matrices $\mathbf{X}_d$ and $\tilde{\mathbf{Y}}_d$ the eigenproblem (15) can be written in the form

$$\mathbf{X}_d^T\tilde{\mathbf{Y}}\tilde{\mathbf{Y}}^T\mathbf{X}_d\mathbf{w} = \lambda\mathbf{w}$$

---

[2] It is considered here that $g \leq N$, otherwise the number of score vectors is given by N.

Thus, similar to the previous two-class discrimination the solution of this eigen-problem can be interpreted as the solution of (16) using the among-classes sum-of-squares matrix now computed on deflated matrix $\mathbf{X}_d$.

A natural further step is to project the original, observed data onto the ob-tained weight vector directions and to build a classifier using this new, projected data representation–PLS score vectors. Support vector machines, logistic regres-sion or other methods for classification can be applied on the extracted PLS score vectors.

5 Nonlinear PLS

In many areas of research and industrial situations data can exhibit nonlinear behaviour. Two major approaches to model nonlinear data relations by means of PLS exist.

A) The first group of approaches is based on reformulating the considered linear relation (5) between the score vectors $\mathbf{t}$ and $\mathbf{u}$ by a nonlinear model

$$\mathbf{u} = g(\mathbf{t}) + \mathbf{h} = g(\mathbf{X}, \mathbf{w}) + \mathbf{h}$$

where $g(.)$ represents a continuous function modeling the existing nonlinear re-lation. Again, $\mathbf{h}$ denotes a vector of residuals. Polynomial functions, smoothing splines, artificial neural networks or radial basis function networks have been used to model $g(.)$ [51, 10, 50, 3].[3] The assumption that the score vectors $\mathbf{t}$ and $\mathbf{u}$ are linear projections of the original variables is kept. This leads to the neces-sity of a linearisation of the nonlinear mapping $g(.)$ by means of Taylor series expansions and to the successive iterative update of the weight vectors $\mathbf{w}$ [51, 3].

B) The second approach to nonlinear PLS is based on a mapping of the orig-inal data by means of a nonlinear function to a new representation (data space) where linear PLS is applied. The recently developed theory of kernel-based learn-ing has been also applied to PLS. The nonlinear kernel PLS methodology was proposed for the modeling of relations between sets of observed variables, regres-sion and classification problems [34, 35]. The idea of the kernel PLS approach is based on the mapping of the original $\mathcal{X}$-space data into a high-dimensional feature space $\mathcal{F}$ corresponding to a reproducing kernel Hilbert space [2, 38]

$$\mathbf{x} \in \mathcal{X} \rightarrow \mathbf{\Phi}(\mathbf{x}) \in \mathcal{F}$$

By applying the *kernel trick* the estimation of PLS in a feature space $\mathcal{F}$ reduces to the use of linear algebra as simple as in linear PLS [34]. The kernel trick uses the fact that a value of a dot product between two vectors in $\mathcal{F}$ can be evaluated by the kernel function [2, 38]

$$k(\mathbf{x}, \mathbf{y}) = \mathbf{\Phi}(\mathbf{x})^T \mathbf{\Phi}(\mathbf{y}), \quad \forall\, \mathbf{x}, \mathbf{y} \in \mathcal{X}$$

---

[3] Note that the below described concept of kernel-based learning can also be used for modeling nonlinear relation between $\mathbf{t}$ and $\mathbf{u}$. An example would be a support vector regression model for $g(.)$ [38].

Define the Gram matrix $\mathbf{K}$ of the cross dot products between all mapped input data points, that is, $\mathbf{K} = \mathbf{\Phi\Phi}^T$, where $\mathbf{\Phi}$ denotes the matrix of mapped $\mathcal{X}$-space data $\{\mathbf{\Phi}(\mathbf{x}_i) \in \mathcal{F}\}_{i=1}^n$. The kernel trick implies that the elements i, j of $\mathbf{K}$ are equal to the values of the kernel function $k(\mathbf{x}_i, \mathbf{x}_j)$. Now, consider a modified version of the NIPALS algorithm where we merge steps 1 and 3 and we scale to unit norm vectors $\mathbf{t}$ and $\mathbf{u}$ instead of the vectors $\mathbf{w}$ and $\mathbf{c}$. We obtain the kernel form of the NIPALS algorithm [34, 20][4]

$$
\begin{aligned}
&1)\ \mathbf{t} = \mathbf{\Phi\Phi}^T\mathbf{u} = \mathbf{Ku} \qquad\qquad &4)\ \mathbf{u} = \mathbf{Yc} \\
&2)\ \|\mathbf{t}\| \rightarrow 1 &5)\ \|\mathbf{u}\| \rightarrow 1 \\
&3)\ \mathbf{c} = \mathbf{Y}^T\mathbf{t}
\end{aligned}
$$

Note that steps 3 and 4 can be further merged which may become useful in applications where an analogous kernel mapping of the $\mathcal{Y}$-space is considered. The kernel PLS approach has been proved to be competitive with the other kernel classification and regression approaches like SVM, kernel RR or kernel FDA [38, 37].

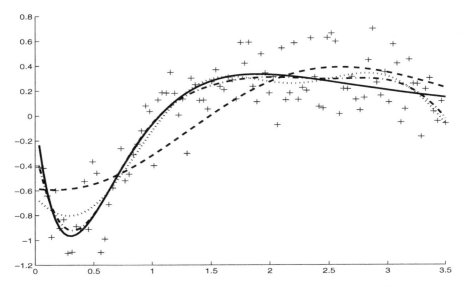

Fig. 3. An example of kernel PLS regression. The generated function $z(.)$ is shown as a solid line. Plus markers represent noisy representation of $z(.)$ used as training output points in kernel PLS regression. Kernel PLS regression using the first one, four and eight score vectors is shown as a dashed, dotted and dash-dotted line, respectively.

When both A) and B) approaches are compared it is difficult to define the favourable methodology. While the kernel PLS approach is easily implementable, computationally less demanding and capable to model difficult nonlinear relations, a loss of the interpretability of the results with respect to the original

---

[4] In the case of the one-dimensional $\mathcal{Y}$-space computationally more efficient kernel PLS algorithms have been proposed in [35, 27].

data limits its use in some applications. On the other hand it is not difficult to construct data situations where the first approach of keeping latent variables as linear projections of the original data may not be adequate. In practice a researcher needs to decide about the adequacy of using a particular approach based on the problem in hands and requirements like simplicity of the solution and implementation or interpretation of the results.

In Figure 3 an example of kernel PLS regression is depicted. We generated one hundred uniformly spaced samples in the range $[0, 3.5]$ and computed the corresponding values of the function [42]

$$z(x) = 4.26(\exp{(-x)} - 4\exp{(-2x)} + 3\exp{(-3x)})$$

Additional one hundred Gaussian distributed samples with zero-mean and variance equal to 0.04 representing noise were generated and added to the computed values. The values of noisy $z(.)$ function were subsequently centered. The Gaussian kernel function $k(\mathbf{x}, \mathbf{y}) = exp(-\frac{\|\mathbf{x}-\mathbf{y}\|^2}{h})$ with the width h equal to 1.8 was used.

6   Conclusions

PLS has been proven to be a very powerful versatile data analytical tool applied in many areas of research and industrial applications. Computational and implementation simplicity of PLS is a strong aspect of the approach which favours PLS to be used as a first step to understand the existing relations and to analyse real world data. The PLS method projects original data onto a more compact space of latent variables. Among many advantages of such an approach is the ability to analyse the importance of individual observed variables potentially leading to the deletion of unimportant variables. This mainly occurs in the case of an experimental design where many insignificant terms are measured. In such situations PLS can guide the practitioner into more compact experimental settings with a significant cost reduction and without a high risk associated with the "blind" variables deletion. Examples of this aspect of PLS are experiments on finger movement detection and cognitive fatigue prediction where a significant reduction of the EEG recording electrodes have been achieved without the loss of classification accuracy of the considered PLS models [35, 43]. Further important aspect of PLS is the ability to visualise high-dimensional data through the set of extracted latent variables. The diagnostic PLS tools based on score and loadings plots allows to better understand data structure, observe existing relations among data sets but also to detect outliers in the measured data.

Successful application of PLS on regression problems associated with many real world data have also attracted attention of statisticians to this method. Although PLS regression is still considered as a method or algorithm rather than a rigorous statistical model, recent advances in understanding of shrinkage properties of PLS regression helped to connect PLS regression with other, in statistical community better understood, shrinkage regression methods like PCR or RR. Moreover, these studies have shown very competitive behaviour of PLS regression in comparison to the other shrinkage regression methods. We believe that further research will reveal additional aspects of PLS regression and will help to better theoretically define structures of data and regression problems where the use of PLS will become beneficial in comparison to the other methods.

Two major approaches of constructing nonlinear PLS have been mentioned. Among other nonlinear versions of PLS, kernel PLS represents an elegant way of dealing with nonlinear aspects of measured data. This method keeps computational and implementation simplicity of linear PLS while providing a powerful modeling, regression, discrimination or classification tool. Kernel PLS approach has been proven to be competitive with the other state-of-the-art kernel-based regression and classification methods.

Connections between PCA, (regularised) CCA and PLS have been highlighted (see [6] for detailed comparison). Understanding of these connections should help to design new algorithms by combining good properties of individual methods and thus resulting in more powerful machine learning tools.

References

1. T. Almøy. A simulation study on comparison of prediction models when only a few components are relevant. *Computational Statistics and Data Analysis*, 21:87–107, 1996.
2. N. Aronszajn. Theory of reproducing kernels. *Transactions of the American Mathematical Society*, 68:337–404, 1950.
3. G. Baffi, E.B. Martin, and A.J. Morris. Non-linear projection to latent structures revisited (the neural network PLS algorithm). *Computers Chemical Engineering*, 23:1293–1307, 1999.
4. M. Barker and W.S. Rayens. Partial least squares for discrimination. *Journal of Chemometrics*, 17:166–173, 2003.
5. M.S. Bartlett. Further aspects of the theory of multiple regression. In *Proceedings of the Cambridge Philosophical Society*, volume 34, pages 33–40, 1938.
6. T. De Bie, N. Cristianini, and R. Rosipal. Eigenproblems in Pattern Recognition. In E. Bayro-Corrochano, editor, *Handbook of Geometric Computing: Applications in Pattern Recognition, Computer Vision, Neuralcomputing, and Robotics*, pages 129–170. Springer, 2005.
7. N.A. Butler and M.C. Denham. The peculiar shrinkage properties of partial least squares regression. *Journal of the Royal Statistical Society: B*, 62:585–593, 2000.
8. S. de Jong. SIMPLS: an alternative approach to partial least squares regression. *Chemometrics and Intelligent Laboratory Systems*, 18:251–263, 1993.
9. S. de Jong, B.M. Wise, and N.L. Ricker. Canonical partial least squares and continuum power regression. *Journal of Chemometrics*, 15:85–100, 2001.
10. I.E. Frank. A nonlinear PLS model. *Chemolab*, 8:109–119, 1990.
11. I.E. Frank and J.H. Friedman. A Statistical View of Some Chemometrics Regression Tools. *Technometrics*, 35:109–147, 1993.
12. C. Goutis. Partial least squares yields shrinkage estimators. *The Annals of Statistics*, 24:816–824, 1996.
13. T. Hastie, R. Tibshirani, and J. Friedman. *The Elements of Statistical Learning*. Springer, 2001.

14. I.S. Helland. Some theoretical aspects of partial least squares regression. *Chemometrics and Intelligent Laboratory Systems*, 58:97–107, 1999.
15. M. Hestenes and E. Stiefel. Methods for conjugate gradients for solving linear systems. *Journal of Research of the National Bureau of Standards*, 49:409–436, 1952.
16. A.E. Hoerl and R.W. Kennard. Ridge regression: bias estimation for nonorthogonal problems. *Technometrics*, 12:55–67, 1970.
17. A. Höskuldsson. PLS Regression Methods. *Journal of Chemometrics*, 2:211–228, 1988.
18. J.S. Hulland. Use of partial least squares (PLS) in strategic management research: A review of four recent studies. *Strategic Management Journal*, 20:195–204, 1999.
19. C. Lanczos. An iteration method for the solution of the eigenvalue problem of linear differential and integral operators. *Journal of Research of the National Bureau of Standards*, 45:225–280, 1950.
20. P.J. Lewi. Pattern recognition, reflection from a chemometric point of view. *Chemometrics and Intelligent Laboratory Systems*, 28:23–33, 1995.
21. O.C. Lingjærde and N. Christophersen. Shrinkage Structure of Partial Least Squares. *Scandinavian Journal of Statistics*, 27:459–473, 2000.
22. N.J. Lobaugh, R. West, and A.R. McIntosh. Spatiotemporal analysis of experimental differences in event-related potential data with partial least squares. *Psychophysiology*, 38:517–530, 2001.
23. R. Manne. Analysis of Two Partial-Least-Squares Algorithms for Multivariate Calibration. *Chemometrics and Intelligent Laboratory Systems*, 2:187–197, 1987.
24. K.V. Mardia, J.T. Kent, and J.M. Bibby. *Multivariate Analysis*. Academic Press, 1997.
25. M. Martens and H. Martens. Partial Least Squares Regression. In J.R. Piggott, editor, *Statistical Procedures in Food Research*, pages 293–359. Elsevier Applied Science, London, 1986.
26. W.F. Massy. Principal components regression in exploratory statistical research. *Journal of the American Statistical Association*, 60:234–256, 1965.
27. M. Momma. Efficient Computations via Scalable Sparse Kernel Partial Least Squares and Boosted Latent Features. In *Proceedings of SIGKDD International Conference on Knowledge and Data Mining*, pages 654–659, Chicago, IL, 2005.
28. D.V. Nguyen and D.M. Rocke. Tumor classification by partial least squares using microarray gene expression data. *Bioinformatics*, 18:39–50, 2002.
29. J. Nilsson, S. de Jong, and A.K. Smilde. Multiway Calibration in 3D QSAR. *Journal of Chemometrics*, 11:511–524, 1997.
30. B. Parlett. *The symmetric eigenvalue problem*. SIAM, 1998.
31. A. Phatak and F. de Hoog. Exploiting the connection between PLS, Lanczos, and conjugate gradients: Alternative proofs of some properties of PLS. *Journal of Chemometrics*, 16:361–367, 2003.
32. A. Phatak, P.M. Rilley, and A. Penlidis. The asymptotic variance of the univariate PLS estimator. *Linear Algebra and its Applications*, 354:245–253, 2002.
33. S. Rännar, F. Lindgren, P. Geladi, and S. Wold. A PLS kernel algorithm for data sets with many variables and fewer objects. Part 1: Theory and algorithm. *Chemometrics and Intelligent Laboratory Systems*, 8:111–125, 1994.
34. R. Rosipal and L.J. Trejo. Kernel Partial Least Squares Regression in Reproducing Kernel Hilbert Space. *Journal of Machine Learning Research*, 2:97–123, 2001.
35. R. Rosipal, L.J. Trejo, and B. Matthews. Kernel PLS-SVC for Linear and Nonlinear Classification. In *Proceedings of the Twentieth International Conference on Machine Learning*, pages 640–647, Washington, DC, 2003.

36. P.D. Sampson, A. P. Streissguth, H.M. Barr, and F.L. Bookstein. Neurobehavioral effects of prenatal alcohol: Part II. Partial Least Squares analysis. *Neurotoxicology and tetralogy*, 11:477–491, 1989.

37. C. Saunders, A. Gammerman, and V. Vovk. Ridge Regression Learning Algorithm in Dual Variables. In *Proceedings of the 15th International Conference on Machine Learning*, pages 515–521, Madison, WI, 1998.

38. B. Schölkopf and A. J. Smola. *Learning with Kernels – Support Vector Machines, Regularization, Optimization and Beyond*. The MIT Press, 2002.

39. A. Smilde, R. Bro, and P. Geladi. *Multi-way Analysis: Applications in the Chemical Sciences*. Wiley, 2004.

40. M. Stone and R.J. Brooks. Continuum Regression: Cross-validated Sequentially Constructed Prediction Embracing Ordinary Least Squares, Partial Least Squares and Principal Components Regression. *Journal of the Royal Statistical Society: B*, 52:237–269, 1990.

41. H. D. Vinod. Canonical ridge and econometrics of joint production. *Journal of Econometrics*, 4:147–166, 1976.

42. G. Wahba. *Splines Models of Observational Data*, volume 59 of *Series in Applied Mathematics*. SIAM, Philadelphia, 1990.

43. J. Wallerius, L.J. Trejo, R. Matthew, R. Rosipal, and J.A. Caldwell. Robust feature extraction and classification of EEG spectra for real-time classification of cognitive state. In *Proceedings of 11th International Conference on Human Computer Interaction*, Las Vegas, NV, 2005.

44. L.E. Wangen and B.R. Kowalsky. A multiblock partial least squares algorithm for investigating complex chemical systems. *Journal of Chemometrics*, 3:3–20, 1989.

45. J.A. Wegelin. A survey of Partial Least Squares (PLS) methods, with emphasis on the two-block case. Technical report, Department of Statistics, University of Washington, Seattle, 2000.

46. J. Westerhuis, T. Kourti, and J. MacGregor. Analysis of multiblock and hierarchical PCA and PLS models. *Journal of Chemometrics*, 12:301–321, 1998.

47. H. Wold. Path models with latent variables: The NIPALS approach. In H.M. Blalock et al., editor, *Quantitative Sociology: International perspectives on mathematical and statistical model building*, pages 307–357. Academic Press, 1975.

48. H. Wold. Soft modeling: the basic design and some extensions. In J.-K. Jöreskog and H. Wold, editor, *Systems Under Indirect Observation*, volume 2, pages 1–53. North Holland, Amsterdam, 1982.

49. H. Wold. Partial least squares. In S. Kotz and N.L. Johnson, editors, *"Encyclopedia of the Statistical Sciences"*, volume 6, pages 581–591. John Wiley & Sons, 1985.

50. S. Wold. Nonlinear partial least squares modeling II, Spline inner relation. *Chemolab*, 14:71–84, 1992.

51. S. Wold, N. Kettaneh-Wold, and B. Skagerberg. Nonlinear PLS modelling. *Chemometrics and Intelligent Laboratory Systems*, 7:53–65, 1989.

52. S. Wold, H. Ruhe, H. Wold, and W.J. Dunn III. The collinearity problem in linear regression. The partial least squares (PLS) approach to generalized inverse. *SIAM Journal of Scientific and Statistical Computations*, 5:735–743, 1984.

53. K.J. Worsley. An overview and some new developments in the statistical analysis of PET and fMRI data. *Human Brain Mapping*, 5:254–258, 1997.

Random Projection, Margins, Kernels, and Feature-Selection

Avrim Blum

Department of Computer Science,
Carnegie Mellon University, Pittsburgh, PA 15213-3891

Abstract. Random projection is a simple technique that has had a
number of applications in algorithm design. In the context of machine
learning, it can provide insight into questions such as "why is a learning
problem easier if data is separable by a large margin?" and "in what sense
is choosing a kernel much like choosing a set of features?" This talk is
intended to provide an introduction to random projection and to survey
some simple learning algorithms and other applications to learning based
on it. I will also discuss how, given a kernel as a black-box function, we
can use various forms of random projection to extract an explicit small
feature space that captures much of what the kernel is doing. This talk
is based in large part on work in [BB05, BBV04] joint with Nina Balcan
and Santosh Vempala.

1 Introduction

Random projection is a technique that has found substantial use in the area
of algorithm design (especially *approximation algorithms*), by allowing one to
substantially reduce dimensionality of a problem while still retaining a significant
degree of problem structure. In particular, given n points in Euclidean space
(of any dimension but which we can think of as R^n), we can project these
points down to a random d-dimensional subspace for $d \ll n$, with the following
outcomes:

1. If $d = \omega(\frac{1}{\gamma^2} \log n)$ then Johnson-Lindenstrauss type results (described below)
   imply that with high probability, relative distances and angles between all
   pairs of points are approximately preserved up to $1 \pm \gamma$.
2. If $d = 1$ (i.e., we project points onto a random line) we can often still get
   something useful.

Projections of the first type have had a number of uses including fast approximate nearest-neighbor algorithms [IM98, EK00] and approximate clustering algorithms [Sch00] among others. Projections of the second type are often used
for "rounding" a semidefinite-programming relaxation, such as for the Max-CUT
problem [GW95], and have been used for various graph-layout problems [Vem98].

   The purpose of this survey is to describe some ways that this technique can
be used (either practically, or for providing insight) in the context of machine

C. Saunders et al. (Eds.): SLSFS 2005, LNCS 3940, pp. 52–68, 2006.
© Springer-Verlag Berlin Heidelberg 2006

learning. In particular, random projection can provide a simple way to see why data that is separable by a large *margin* is easy for learning even if data lies in a high-dimensional space (e.g., because such data can be randomly projected down to a low dimensional space without affecting separability, and therefore it is "really" a low-dimensional problem after all). It can also suggest some especially simple algorithms. In addition, random projection (of various types) can be used to provide an interesting perspective on *kernel* functions, and also provide a mechanism for converting a kernel function into an explicit feature space.

The use of Johnson-Lindenstrauss type results in the context of learning was first proposed by Arriaga and Vempala [AV99], and a number of uses of random projection in learning are discussed in [Vem04]. Experimental work on using random projection has been performed in [FM03, GBN05, Das00]. This survey, in addition to background material, focuses primarily on work in [BB05, BBV04]. Except in a few places (e.g., Theorem 1, Lemma 1) we give only sketches and basic intuition for proofs, leaving the full proofs to the papers cited.

1.1   The Setting

We are considering the standard PAC-style setting of supervised learning from i.i.d. data. Specifically, we assume that examples are given to us according to some probability distribution D over an instance space X and labeled by some unknown target function $c : X \to \{-1, +1\}$. We use $P = (D, c)$ to denote the combined distribution over labeled examples. Given some sample S of labeled training examples (each drawn independently from D and labeled by c), our objective is to come up with a hypothesis h with low true error: that is, we want $\mathrm{Pr}_{x \sim D}(h(x) \neq c(x))$ to be low. In the discussion below, by a "learning problem" we mean a distribution $P = (D, c)$ over labeled examples.

In the first part of this survey (Sections 2 and 3), we will think of the input space X as Euclidean space, like R^n. In the second part (Section 4), we will discuss kernel functions, in which case one should think of X as just some abstract space, and a kernel function $K : X \times X \to [-1, 1]$ is then some function that provides a measure of similarity between two input points. Formally, one requires for a legal kernel K that there exist some implicit function ϕ mapping X into a (possibly very high-dimensional) space, such that $K(x, y) = \phi(x) \cdot \phi(y)$. In fact, one interesting property of some of the results we discuss is that they make sense to apply even if K is just an arbitrary similarity function, and not a "legal" kernel, though the theorems make sense only if such a ϕ exists. Extensions of this framework to more general similarity functions are given in [BB06].

Definition 1. *We say that a set S of labeled examples is* **linearly separable** **by margin** γ *if there exists a unit-length vector w such that:*

$$\min_{(x, \ell) \in S} [\ell(w \cdot x) / ||x||] \geq \gamma.$$

That is, the separator $w \cdot x \geq 0$ has margin γ if every labeled example in S is correctly classified and furthermore the cosine of the angle between w and x has magnitude at least γ.[1] For simplicity, we are only considering separators that pass through the origin, though results we discuss can be adapted to the general case as well.

We can similarly talk in terms of the distribution P rather than a sample S.

Definition 2. *We say that* P *is* **linearly separable by margin** γ *if there exists a unit-length vector* w *such that:*

$$\Pr_{(x,\ell)\sim P}[\ell(w\cdot x)/||x|| < \gamma] = 0,$$

and we say that P *is* **separable with error** α *at margin* γ *if there exists a unit-length vector* w *such that:*

$$\Pr_{(x,\ell)\sim P}[\ell(w\cdot x)/||x|| < \gamma] \leq \alpha.$$

A powerful theoretical result in machine learning is that if a learning problem is linearly separable by a large margin γ, then that makes the problem "easy" in the sense that to achieve good generalization one needs only a number of examples that depends (polynomially) on $1/\gamma$, with no dependence on the dimension of the ambient space X that examples lie in. In fact, two results of this form are:

1. The classic Perceptron Convergence Theorem that the Perceptron Algorithm makes at most $1/\gamma^2$ mistakes on any sequence of examples separable by margin γ [Blo62, Nov62, MP69]. Thus, if the Perceptron algorithm is run on a sample of size $1/(\epsilon\gamma^2)$, the expected error rate of its hypothesis at a random point in time is at most ϵ. (For further results of this form, see [Lit89, FS99]).

2. The more recent margin bounds of [STBWA98, BST99] that state that $|S| = O(\frac{1}{\epsilon}[\frac{1}{\gamma^2}\log^2(\frac{1}{\gamma\epsilon}) + \log\frac{1}{\delta}])$ is sufficient so that with high probability, *any* linear separator of S with margin γ has true error at most ϵ. Thus, this provides a sample complexity bound that applies to any algorithm that finds large-margin separators.

In the next two sections, we give two more ways of seeing why having a large margin makes a learning problem easy, both based on the idea of random projection.

2   An Extremely Simple Learning Algorithm

In this section, we show how the idea of random projection can be used to get an extremely simple algorithm (almost embarrassingly so) for *weak*-learning, with

---

[1] Often margin is defined without normalizing by the length of the examples, though in that case the "γ^2" term in sample complexity bounds becomes "γ^2/R^2", where R is the maximum $||x||$ over $x \in S$. Technically, normalizing produces a stronger bound because we are taking the minimum of a ratio, rather than the ratio of a minimum to a maximum.

error rate $1/2 - \gamma/4$, whenever a learning problem is linearly separable by some margin γ. This can then be plugged into Boosting [Sch90, FS97], to achieve strong-learning. This material is taken from [BB05].

In particular, the algorithm is as follows.

Algorithm 1 (Weak-learning a linear separator)

1. *Pick a random linear separator. Specifically, choose a random unit-length vector h and consider the separator $h \cdot x \geq 0$. (Equivalently, project data to the 1-dimensional line spanned by h and consider labeling positive numbers as positive and negative numbers as negative.)*
2. *Evaluate the error of the separator selected in Step 1. If the error is at most $1/2 - \gamma/4$ then halt with success, else go back to 1.*

Theorem 1. *If P is separable by margin γ then a random linear separator will have error at most $\frac{1}{2} - \gamma/4$ with probability $\Omega(\gamma)$.*

In particular, Theorem 1 implies that the above algorithm will in expectation repeat only $O(1/\gamma)$ times before halting.[2]

Proof. Consider a (positive) example x such that $x \cdot w^*/\|x\| \geq \gamma$. The angle between x and w^* is then some value $\alpha \leq \pi/2 - \gamma$. Now, a random vector h, when projected onto the 2-dimensional plane defined by x and w^*, looks like a random vector in this plane. Therefore, we have (see Figure 1):

$$\Pr_h(h \cdot x \leq 0 | h \cdot w^* \geq 0) = \alpha/\pi \leq 1/2 - \gamma/\pi.$$

Similarly, for a negative example x, for which $x \cdot w^*/\|x\| \leq -\gamma$, we have:

$$\Pr_h(h \cdot x \geq 0 | h \cdot w^* \geq 0) \leq 1/2 - \gamma/\pi.$$

Therefore, if we define $h(x)$ to be the classifier defined by $h \cdot x \geq 0$, we have:

$$\mathbf{E}[err(h) | h \cdot w^* \geq 0] \leq 1/2 - \gamma/\pi.$$

Finally, since the error rate of any hypothesis is bounded between 0 and 1, and a random vector h has a $1/2$ chance of satisfying $h \cdot w^* \geq 0$, it must be the case that:

$$\Pr_h[err(h) \leq 1/2 - \gamma/4] = \Omega(\gamma). \qquad \square$$

---

[2] For simplicity, we have presented Algorithm 1 as if it can exactly evaluate the true error of its chosen hypothesis in Step 2. Technically, we should change Step 2 to talk about empirical error, using an intermediate value such as $1/2 - \gamma/6$. In that case, sample size $O(\frac{1}{\gamma^2} \log \frac{1}{\gamma})$ is sufficient to be able to run Algorithm 1 for $O(1/\gamma)$ repetitions, and evaluate the error rate of each hypothesis produced to sufficient precision.

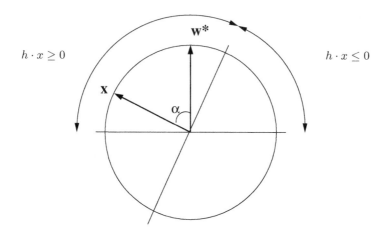

Fig. 1. For random h, conditioned on $h \cdot w^* \geq 0$, we have $\Pr(h \cdot x \leq 0) = \alpha/\pi$ and $\Pr(h \cdot x \geq 0) = 1 - \alpha/\pi$

3   The Johnson-Lindenstrauss Lemma

The Johnson-Lindenstrauss Lemma [JL84, DG02, IM98] states that given a set S of points in R^n, if we perform an orthogonal projection of those points onto a random d-dimensional subspace, then $d = O(\frac{1}{\gamma^2} \log |S|)$ is sufficient so that with high probability all pairwise distances are preserved up to $1 \pm \gamma$ (up to scaling). Conceptually, one can think of a random projection as first applying a random rotation to R^n and then reading off the first d coordinates. In fact, a number of different forms of "random projection" are known to work (including some that are especially efficient to perform computationally, considered in [Ach03, AV99]). In particular, if we think of performing the projection via multiplying all points, viewed as row-vectors of length n, by an $n \times d$ matrix A, then several methods for selecting A that provide the desired result are:

1. Choosing its columns to be d random orthogonal unit-length vectors (a true random orthogonal projection).
2. Choosing each entry in A independently from a standard Gaussian (so the projection can be viewed as selecting d vectors $u_1, u_2, \ldots, u_d$ from a spherical gaussian and mapping a point p to $(p \cdot u_1, \ldots, p \cdot u_d)$.
3. Choosing each entry in A to be 1 or -1 independently at random.

Some especially nice proofs for the Johnson-Lindenstrauss Lemma are given by Indyk and Motwani [IM98] and Dasgupta and Gupta [DG02]. Here, we just give the basic structure and intuition for the argument. In particular, consider two points p_i and p_j in the input and their difference $v_{ij} = p_i - p_j$. So, we are interested in the length of $v_{ij}A$. Fixing v_{ij}, let us think of each of the d coordinates $y_1, \ldots, y_d$ in the vector $y = v_{ij}A$ as random variables (over the choice of A). Then, in each form of projection above, these d random variables are nearly independent (in fact, in forms (2) and (3) they are completely independent random variables). This allows us to use a Chernoff-style bound to argue that $d = O(\frac{1}{\gamma^2} \log \frac{1}{\delta})$ is sufficient so that with probability $1 - \delta$, $y_1^2 + \ldots + y_d^2$ will be within $1 \pm \gamma$ of its expectation. This in turn implies that the length of y is within $1 \pm \gamma$ of its expectation. Finally, using $\delta = o(1/n^2)$ we have by the union bound that with high probability this is satisfied simultaneously for all pairs of points p_i, p_j in the input.

Formally, here is a convenient form of the Johnson-Lindenstrauss Lemma given in [AV99]. Let $N(0, 1)$ denote the standard Normal distribution with mean 0 and variance 1, and $U(-1, 1)$ denote the distribution that has probability $1/2$ on -1 and probability $1/2$ on 1.

Theorem 2 (Neuronal RP [AV99]). *Let $u, v \in R^n$. Let $u' = \frac{1}{\sqrt{d}}uA$ and $v' = \frac{1}{\sqrt{d}}vA$ where A is a $n \times d$ random matrix whose entries are chosen independently from either $N(0, 1)$ or $U(-1, 1)$. Then,*

$$\Pr_A \left[(1 - \gamma)||u - v||^2 \leq ||u' - v'||^2 \leq (1 + \gamma)||u - v||^2 \right] \geq 1 - 2e^{-(\gamma^2 - \gamma^3)\frac{d}{4}}.$$

Theorem 2 suggests a natural learning algorithm: first randomly project data into a lower dimensional space, and then run some other algorithm in that space, taking advantage of the speedup produced by working over fewer dimensions. Theoretical results for some algorithms of this form are given in [AV99], and experimental results are given in [FM03, GBN05, Das00].

3.1 The Johnson-Lindenstrauss Lemma and Margins

The Johnson-Lindenstrauss lemma provides a particularly intuitive way to see why one should be able to generalize well from only a small amount of training data when a learning problem is separable by a large margin. In particular, imagine a set S of data in some high-dimensional space, and suppose that we randomly project the data down to R^d. By the Johnson-Lindenstrauss Lemma, $d = O(\gamma^{-2} \log |S|)$ is sufficient so that with high probability, all *angles* between points (viewed as vectors) change by at most $\pm\gamma/2$.[3] In particular, consider projecting all points in S *and* the target vector w^*; if initially data was separable by margin γ, then after projection, since angles with w^* have changed by at most $\gamma/2$, the data is still separable (and in fact separable by margin $\gamma/2$). Thus, this means our problem was in some sense really only a d-dimensional problem after all. Moreover, if we replace the "$\log |S|$" term in the bound for d with "$\log \frac{1}{\epsilon}$", then we can use Theorem 2 to get that with high probability at least a $1 - \epsilon$ fraction of S will be separable. Formally, talking in terms of the true distribution P, one can state the following theorem. (Proofs appear in, e.g., [AV99, BBV04].)

---

[3] The Johnson-Lindenstrauss Lemma talks about relative *distances* being approximately preserved, but it is a straightforward calculation to show that this implies angles must be approximately preserved as well.

Theorem 3. *If P is linearly separable by margin γ, then $d = O\left(\frac{1}{\gamma^2}\log(\frac{1}{\varepsilon\delta})\right)$ is sufficient so that with probability at least $1 - \delta$, a random projection down to R^d will be linearly separable with error at most ε at margin $\gamma/2$.*

So, Theorem 3 can be viewed as stating that a learning problem separable by margin γ is really only an "$O(1/\gamma^2)$-dimensional problem" after all.

4   Random Projection, Kernel Functions, and Feature Selection

4.1   Introduction

Kernel functions [BGV92, CV95, FS99, MMR⁺01, STBWA98, Vap98] have become a powerful tool in Machine Learning. A kernel function can be viewed as allowing one to implicitly map data into a high-dimensional space and to perform certain operations there without paying a high price computationally. Furthermore, margin bounds imply that if the learning problem has a large margin linear separator in that space, then one can avoid paying a high price in terms of sample size as well.

Combining kernel functions with the Johnson-Lindenstrauss Lemma (in particular, Theorem 3 above), we have that if a learning problem indeed has the large margin property under kernel $K(x, y) = \phi(x) \cdot \phi(y)$, then a *random* linear projection of the "ϕ-space" down to a *low* dimensional space approximately preserves linear separability. This means that for any kernel K under which the learning problem is linearly separable by margin γ in the ϕ-space, we can, in principle, think of K as mapping the input space X into an $\tilde{O}(1/\gamma^2)$-dimensional space, in essence serving as a method for representing the data in a new (and not too large) feature space.

The question we now consider is whether, given kernel K as a black-box function, we can in fact produce such a mapping efficiently. The problem with the above observation is that it requires explicitly computing the function $\phi(x)$. Since for a given kernel K, the dimensionality of the ϕ-space might be quite large, this is not efficient.[4] Instead, what we would like is an efficient procedure that given $K(.,.)$ as a black-box program, produces a mapping with the desired properties using running time that depends (polynomially) only on $1/\gamma$ and the time to compute the kernel function K, with no dependence on the dimensionality of the ϕ-space. This would mean we can effectively convert a kernel K that is good for some learning problem into an explicit set of features, without a need for "kernelizing" our learning algorithm. In this section, we describe several methods for doing so; this work is taken from [BBV04].

Specifically, we will show the following. Given black-box access to a kernel function $K(x, y)$, access to unlabeled examples from distribution D, and parameters γ, ε, and δ, we can in polynomial time construct a mapping $F : X \to R^d$

---

[4] In addition, it is not totally clear how to apply Theorem 2 if the dimension of the ϕ-space is infinite.

(i.e., to a set of d real-valued features) where $d = O\left(\frac{1}{\gamma^2} \log \frac{1}{\varepsilon\delta}\right)$, such that if the target concept indeed has margin γ in the ϕ-space, then with probability $1 - \delta$ (over randomization in our choice of mapping function), the induced distribution in R^d is separable with error $\leq \varepsilon$. In fact, not only will the data in R^d be separable, but it will be separable with a margin $\gamma' = \Omega(\gamma)$. (If the original learning problem was separable with error α at margin γ then the induced distribution is separable with error $\alpha + \epsilon$ at margin γ'.)

To give a feel of what such a mapping might look like, suppose we are willing to use dimension $d = O(\frac{1}{\varepsilon}[\frac{1}{\gamma^2} + \ln \frac{1}{\delta}])$ (so this is linear in $1/\varepsilon$ rather than logarithmic) and we are not concerned with preserving margins and only want approximate separability. Then we show the following especially simple procedure suffices. Just draw a random sample of d unlabeled points $x_1, \ldots, x_d$ from D and define $F(x) = (K(x, x_1), \ldots, K(x, x_d))$.[5] That is, if we think of K not so much as an implicit mapping into a high-dimensional space but just as a similarity function over examples, what we are doing is drawing d "reference" points and then defining the ith feature of x to be its similarity with reference point i. Corollary 1 (in Section 4.3 below) shows that under the assumption that the target function has margin γ in the ϕ space, with high probability the data will be approximately separable under this mapping. Thus, this gives a particularly simple way of using the kernel and unlabeled data for feature generation.

Given these results, a natural question is whether it might be possible to perform mappings of this type without access to the underlying distribution. In Section 4.6 we show that this is in general *not* possible, given only black-box access (and polynomially-many queries) to an *arbitrary* kernel K. However, it may well be possible for specific standard kernels such as the polynomial kernel or the gaussian kernel.

4.2 Additional Definitions

In analogy to Definition 2, we will say that P is separable by margin γ in the ϕ-space if there exists a unit-length vector w in the ϕ-space such that $\Pr_{(x,\ell)\sim P}[\ell(w \cdot \phi(x))/||\phi(x)|| < \gamma] = 0$, and similarly that P is separable with error α at margin γ in the ϕ-space if the above holds with "$= 0$" replaced by "$\leq \alpha$".

For a set of vectors $v_1, v_2, \ldots, v_k$ in Euclidean space, let $\mathrm{span}(v_1, \ldots, v_k)$ denote the span of these vectors: that is, the set of vectors u that can be written as a linear combination $a_1 v_1 + \ldots + a_k v_k$. Also, for a vector u and a subspace Y, let $\mathrm{proj}(u, Y)$ be the orthogonal projection of u down to Y. So, for instance, $\mathrm{proj}(u, \mathrm{span}(v_1, \ldots, v_k))$ is the orthogonal projection of u down to the space spanned by $v_1, \ldots, v_k$. We note that given a set of vectors $v_1, \ldots, v_k$ and the ability to compute dot-products, this projection can be computed efficiently by solving a set of linear equalities.

---

[5] In contrast, the Johnson-Lindenstrauss Lemma as presented in Theorem 2 would draw d Gaussian (or uniform $\{-1, +1\}$) random points $r_1, \ldots, r_d$ in the ϕ-space and define $F(x) = (\phi(x) \cdot r_1, \ldots, \phi(x) \cdot r_d)$.

4.3   Two Simple Mappings

Our goal is a procedure that given black-box access to a kernel function $K(.,.)$, unlabeled examples from distribution D, and a margin value γ, produces a mapping $F : X \to R^d$ with the following property: if the target function indeed has margin γ in the ϕ-space, then with high probability F approximately preserves linear separability. In this section, we analyze two methods that both produce a space of dimension $O(\frac{1}{\varepsilon}[\frac{1}{\gamma^2} + \ln \frac{1}{\delta}])$, such that with probability $1 - \delta$ the result is separable with error at most ε. The second of these mappings in fact satisfies a stronger condition that its output will be approximately separable at margin $\gamma/2$ (rather than just approximately separable). This property will allow us to use this mapping as a first step in a better mapping in Section 4.4.

The following lemma is key to our analysis.

Lemma 1. *Consider any distribution over labeled examples in Euclidean space such that there exists a linear separator $w \cdot x = 0$ with margin γ. If we draw*

$$d \geq \frac{8}{\varepsilon} \left[\frac{1}{\gamma^2} + \ln \frac{1}{\delta} \right]$$

examples $z_1, \ldots, z_d$ iid from this distribution, with probability $\geq 1-\delta$, there exists a vector w' in span$(z_1, \ldots, z_d)$ that has error at most ε at margin $\gamma/2$.

Remark 1. Before proving Lemma 1, we remark that a somewhat weaker bound on d can be derived from the machinery of margin bounds. Margin bounds [STBWA98, BST99] tell us that using $d = O(\frac{1}{\varepsilon}[\frac{1}{\gamma^2} \log^2(\frac{1}{\gamma\varepsilon}) + \log \frac{1}{\delta}])$ points, with probability $1-\delta$, *any* separator with margin $\geq \gamma$ over the observed data has true error $\leq \varepsilon$. Thus, the projection of the target function w into the space spanned by the observed data will have true error $\leq \varepsilon$ as well. (Projecting w into this space maintains the value of $w \cdot z_i$, while possibly shrinking the vector w, which can only increase the margin over the observed data.) The only technical issue is that we want as a conclusion for the separator not only to have a low error rate over the distribution, but also to have a large margin. However, this can be obtained from the double-sample argument used in [STBWA98, BST99] by using a $\gamma/4$-cover instead of a $\gamma/2$-cover. Margin bounds, however, are a bit of an overkill for our needs, since we are only asking for an existential statement (the *existence* of w') and not a universal statement about all separators with large empirical margins. For this reason we are able to get a better bound by a direct argument from first principles.

Proof (Lemma 1). For any set of points S, let $w_{in}(S)$ be the projection of w to span(S), and let $w_{out}(S)$ be the orthogonal portion of w, so that $w = w_{in}(S) + w_{out}(S)$ and $w_{in}(S) \perp w_{out}(S)$. Also, for convenience, assume w and all examples z are unit-length vectors (since we have defined margins in terms of angles, we can do this without loss of generality). Now, let us make the following definitions. Say that $w_{out}(S)$ is *large* if $\Pr_z(|w_{out}(S) \cdot z| > \gamma/2) \geq \varepsilon$, and otherwise say that $w_{out}(S)$ is *small*. Notice that if $w_{out}(S)$ is small, we are done, because $w \cdot z = (w_{in}(S) \cdot z) + (w_{out}(S) \cdot z)$, which means that $w_{in}(S)$ has the properties we want. That is, there is at most an ε probability mass of points z whose dot-product with w and $w_{in}(S)$ differ by more than $\gamma/2$. So, we need only to consider what happens when $w_{out}(S)$ is large.

The crux of the proof now is that if $w_{out}(S)$ is large, this means that a new random point z has at least an ε chance of significantly improving the set S. Specifically, consider z such that $|w_{out}(S) \cdot z| > \gamma/2$. Let $z_{in}(S)$ be the projection of z to span(S), let $z_{out}(S) = z - z_{in}(S)$ be the portion of z orthogonal to span(S), and let $z' = z_{out}(S)/||z_{out}(S)||$. Now, for $S' = S \cup \{z\}$, we have $w_{out}(S') = w_{out}(S) - \text{proj}(w_{out}(S), \text{span}(S')) = w_{out}(S) - (w_{out}(S) \cdot z')z'$, where the last equality holds because $w_{out}(S)$ is orthogonal to span(S) and so its projection onto span(S') is the same as its projection onto z'. Finally, since $w_{out}(S')$ is orthogonal to z' we have $||w_{out}(S')||^2 = ||w_{out}(S)||^2 - |w_{out}(S) \cdot z'|^2$, and since $|w_{out}(S) \cdot z'| \geq |w_{out}(S) \cdot z_{out}(S)| = |w_{out}(S) \cdot z|$, this implies by definition of z that $||w_{out}(S')||^2 < ||w_{out}(S)||^2 - (\gamma/2)^2$.

So, we have a situation where so long as w_{out} is large, each example has at least an ε chance of reducing $||w_{out}||^2$ by at least $\gamma^2/4$, and since $||w||^2 = ||w_{out}(\emptyset)||^2 = 1$, this can happen at most $4/\gamma^2$ times. Chernoff bounds state that a coin of bias ε flipped $n = \frac{8}{\varepsilon}\left[\frac{1}{\gamma^2} + \ln\frac{1}{\delta}\right]$ times will with probability $1 - \delta$ have at least $n\varepsilon/2 \geq 4/\gamma^2$ heads. Together, these imply that with probability at least $1 - \delta$, $w_{out}(S)$ will be small for $|S| \geq \frac{8}{\varepsilon}\left[\frac{1}{\gamma^2} + \ln\frac{1}{\delta}\right]$ as desired.   □

Lemma 1 implies that if P is linearly separable with margin γ under K, and we draw $d = \frac{8}{\varepsilon}[\frac{1}{\gamma^2} + \ln\frac{1}{\delta}]$ random unlabeled examples $x_1, \ldots, x_n$ from D, then with probability at least $1 - \delta$ there is a separator w' in the ϕ-space with error rate at most ε that can be written as

$$w' = \alpha_1\phi(x_1) + \ldots + \alpha_d\phi(x_d).$$

Notice that since $w' \cdot \phi(x) = \alpha_1 K(x, x_1) + \ldots + \alpha_d K(x, x_d)$, an immediate implication is that if we simply think of $K(x, x_i)$ as the ith "feature" of x — that is, if we define $F_1(x) = (K(x, x_1), \ldots, K(x, x_d))$ — then with high probability the vector $(\alpha_1, \ldots, \alpha_d)$ is an approximate linear separator of $F_1(P)$. So, the kernel and distribution together give us a particularly simple way of performing feature generation that preserves (approximate) separability. Formally, we have the following.

Corollary 1. *If P has margin γ in the ϕ-space, then with probability $\geq 1 - \delta$, if $x_1, \ldots, x_d$ are drawn from D for $d = \frac{8}{\varepsilon}\left[\frac{1}{\gamma^2} + \ln\frac{1}{\delta}\right]$, the mapping*

$$F_1(x) = (K(x, x_1), \ldots, K(x, x_d))$$

produces a distribution $F_1(P)$ on labeled examples in R^d that is linearly separable with error at most ε.

Unfortunately, the above mapping F_1 may not preserve margins because we do not have a good bound on the length of the vector $(\alpha_1, \ldots, \alpha_d)$ defining the separator in the new space, or the length of the examples $F_1(x)$. The key problem is that if many of the $\phi(x_i)$ are very similar, then their associated features $K(x, x_i)$ will be highly correlated. Instead, to preserve margin we want to choose an orthonormal basis of the space spanned by the $\phi(x_i)$: i.e., to do an orthogonal projection of $\phi(x)$ into this space. Specifically, let $S = \{x_1, ..., x_d\}$ be a set of $\frac{8}{\varepsilon}[\frac{1}{\gamma^2} + \ln \frac{1}{\delta}]$ unlabeled examples from D as in Corollary 1. We can then implement the desired orthogonal projection of $\phi(x)$ as follows. Run $K(x, y)$ for all pairs $x, y \in S$, and let $M(S) = (K(x_i, x_j))_{x_i, x_j \in S}$ be the resulting kernel matrix. Now decompose $M(S)$ into $U^T U$, where U is an upper-triangular matrix. Finally, define the mapping $F_2 : X \to R^d$ to be $F_2(x) = F_1(x)U^{-1}$, where F_1 is the mapping of Corollary 1. This is equivalent to an orthogonal projection of $\phi(x)$ into span$(\phi(x_1), \ldots, \phi(x_d))$. Technically, if U is not full rank then we want to use the (Moore-Penrose) pseudoinverse [BIG74] of U in place of U^{-1}.

By Lemma 1, this mapping F_2 maintains approximate separability at margin $\gamma/2$ (See [BBV04] for a full proof):

Theorem 4. *If P has margin γ in the ϕ-space, then with probability $\geq 1 - \delta$, the mapping $F_2 : X \to R^d$ for $d \geq \frac{8}{\varepsilon}\left[\frac{1}{\gamma^2} + \ln \frac{1}{\delta}\right]$ has the property that $F_2(P)$ is linearly separable with error at most ε at margin $\gamma/2$.*

Notice that the running time to compute $F_2(x)$ is polynomial in $1/\gamma, 1/\varepsilon, 1/\delta$ and the time to compute the kernel function K.

4.4   An Improved Mapping

We now describe an improved mapping, in which the dimension d has only a logarithmic, rather than linear, dependence on $1/\varepsilon$. The idea is to perform a two-stage process, composing the mapping from the previous section with an additional Johnson-Lindenstrauss style mapping to reduce dimensionality even further. Thus, this mapping can be thought of as combining two types of random projection: a projection based on points chosen at random from D, and a projection based on choosing points uniformly at random in the intermediate space.

In particular, let $F_2 : X \to R^{d_2}$ be the mapping from Section 4.3 using $\varepsilon/2$ and $\delta/2$ as its error and confidence parameters respectively. Let $\hat{F} : R^{d_2} \to R^{d_3}$ be a random projection as in Theorem 2. Then consider the overall mapping $F_3 : X \to R^{d_3}$ to be $F_3(x) = \hat{F}(F_2(x))$.

We now claim that for $d_2 = O(\frac{1}{\varepsilon}[\frac{1}{\gamma^2} + \ln \frac{1}{\delta}])$ and $d_3 = O(\frac{1}{\gamma^2} \log(\frac{1}{\varepsilon\delta}))$, with high probability, this mapping has the desired properties. The basic argument is that the initial mapping F_2 maintains approximate separability at margin $\gamma/2$ by Lemma 1, and then the second mapping approximately preserves this property by Theorem 2. In particular, we have (see [BBV04] for a full proof):

Theorem 5. *If P has margin γ in the ϕ-space, then with probability at least $1 - \delta$, the mapping $F_3 = \hat{F} \circ F_2 : X \to R^{d_3}$, for values $d_2 = O\left(\frac{1}{\varepsilon}\left[\frac{1}{\gamma^2} + \ln \frac{1}{\delta}\right]\right)$ and $d_3 = O\left(\frac{1}{\gamma^2} \log(\frac{1}{\varepsilon\delta})\right)$, has the property that $F_3(P)$ is linearly separable with error at most ε at margin $\gamma/4$.*

As before, the running time to compute our mappings is polynomial in $1/\gamma$, $1/\varepsilon, 1/\delta$ and the time to compute the kernel function K.

Since the dimension d_3 of the mapping in Theorem 5 is only logarithmic in $1/\varepsilon$, this means that if P is perfectly separable with margin γ in the ϕ-space, we can set ε to be small enough so that with high probability, a sample of size $O(d_3 \log d_3)$ would be perfectly separable. That is, we could use an arbitrary noise-free linear-separator learning algorithm in R^{d_3} to learn the target concept. However, this requires using $d_2 = \tilde{O}(1/\gamma^4)$ (i.e., $\tilde{O}(1/\gamma^4)$ unlabeled examples to construct the mapping).

Corollary 2. *Given $\varepsilon', \delta, \gamma < 1$, if P has margin γ in the ϕ-space, then $\tilde{O}(\frac{1}{\varepsilon'\gamma^4})$ unlabeled examples are sufficient so that with probability $1-\delta$, mapping $F_3 : X \to R^{d_3}$ has the property that $F_3(P)$ is linearly separable with error $o(\varepsilon'/(d_3 \log d_3))$, where $d_3 = O(\frac{1}{\gamma^2} \log \frac{1}{\varepsilon'\gamma\delta})$.*

4.5   A Few Extensions

So far, we have assumed that the distribution P is *perfectly* separable with margin γ in the ϕ-space. Suppose, however, that P is only separable with error α at margin γ. That is, there exists a vector w in the ϕ-space that correctly classifies a $1 - \alpha$ probability mass of examples by margin at least γ, but the remaining α probability mass may be either within the margin or incorrectly classified. In that case, we can apply all the previous results to the $1 - \alpha$ portion of the distribution that is correctly separated by margin γ, and the remaining α probability mass of examples may or may not behave as desired. Thus all preceding results (Lemma 1, Corollary 1, Theorem 4, and Theorem 5) still hold, but with ε replaced by $(1 - \alpha)\varepsilon + \alpha$ in the error rate of the resulting mapping.

Another extension is to the case that the target separator does not pass through the origin: that is, it is of the form $w \cdot \phi(x) \geq \beta$ for some value β. If our kernel function is normalized, so that $||\phi(x)|| = 1$ for all $x \in X$, then all results carry over directly (note that one can normalize any kernel K by defining $\hat{K}(x, x') = K(x, x')/\sqrt{K(x, x)K(x', x')}$). In particular, all our results follow from arguments showing that the cosine of the angle between w and $\phi(x)$ changes by at most ε due to the reduction in dimension. If the kernel is not normalized, then results still carry over if one is willing to divide by the maximum value of $||\phi(x)||$, but we do not know if results carry over if one wishes to be truly translation-independent, say bounding only the radius of the smallest ball enclosing all $\phi(x)$ but not necessarily centered at the origin.

4.6   On the Necessity of Access to D

Our algorithms construct mappings $F : X \to R^d$ using black-box access to the kernel function $K(x, y)$ together with unlabeled examples from the input distribution D. It is natural to ask whether it might be possible to remove the need for access to D. In particular, notice that the mapping resulting from the Johnson-Lindenstrauss lemma has nothing to do with the input distribution:

if we have access to the ϕ-space, then no matter what the distribution is, a random projection down to R^d will approximately preserve the existence of a large-margin separator with high probability.[6] So perhaps such a mapping F can be produced by just computing K on some polynomial number of cleverly-chosen (or uniform random) points in X. (Let us assume X is a "nice" space such as the unit ball or $\{0,1\}^n$ that can be randomly sampled.) In this section, we show this is not possible in general for an arbitrary black-box kernel. This leaves open, however, the case of specific natural kernels.

One way to view the result of this section is as follows. If we define a feature space based on dot-products with uniform or gaussian-random points in the ϕ-space, then we know this will work by the Johnson-Lindenstrauss lemma. However, this requires explicit access to the ϕ-space. Alternatively, using Corollary 1 we can define features based on dot-products with points $\phi(x)$ for $x \in X$, which only requires *implicit* access to the ϕ-space through the kernel. However, this procedure needs to use D to select the points x. What we show here is that such use of D is necessary: if we define features based on points $\phi(x)$ for uniform random $x \in X$, or any other distribution that does not depend on D, then there will exist kernels for which this does not work.

We demonstrate the necessity of access to D as follows. Consider $X = \{0,1\}^n$, let X' be a random subset of $2^{n/2}$ elements of X, and let D be the uniform distribution on X'. For a given target function c, we will define a special ϕ-function ϕ_c such that c is a large margin separator in the ϕ-space under distribution D, but that only the points in X' behave nicely, and points not in X' provide no useful information. Specifically, consider $\phi_c : X \rightarrow R^2$ defined as:

$$\phi_c(x) = \begin{cases} (1,0) & \text{if } x \notin X' \\ (-1/2, \sqrt{3}/2) & \text{if } x \in X' \text{ and } c(x) = 1 \\ (-1/2, -\sqrt{3}/2) & \text{if } x \in X' \text{ and } c(x) = -1 \end{cases}$$

See figure 2. This then induces the kernel:

$$K_c(x,y) = \begin{cases} 1 & \text{if } x, y \notin X' \text{ or } [x, y \in X' \text{ and } c(x) = c(y)] \\ -1/2 & \text{otherwise} \end{cases}$$

Notice that the distribution $P = (D, c)$ over labeled examples has margin $\gamma = \sqrt{3}/2$ in the ϕ-space.

Theorem 6. *Suppose an algorithm makes polynomially many calls to a black-box kernel function over input space $\{0,1\}^n$ and produces a mapping $F : X \rightarrow R^d$ where d is polynomial in n. Then for random X' and random c in the above construction, with high probability $F(P)$ will not even be weakly-separable (even though P has margin $\gamma = \sqrt{3}/2$ in the ϕ-space).*

---

[6] To be clear about the order of quantification, the statement is that for any distribution, a random projection will work with high probability. However, for any given projection, there may exist bad distributions. So, even if we could define a mapping of the sort desired, we would still expect the algorithm to be randomized.

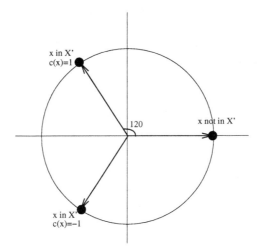

Fig. 2. Function ϕ_c used in lower bound

Proof Sketch: Consider any algorithm with black-box access to K attempting to create a mapping $F : X \rightarrow R^d$. Since X' is a random exponentially-small fraction of X, with high probability all calls made to K when constructing the function F are on inputs not in X'. Let us assume this indeed is the case. This implies that (a) all calls made to K when constructing the function F return the value 1, and (b) at "runtime" when x chosen from D (i.e., when F is used to map training data), even though the function $F(x)$ may itself call $K(x, y)$ for different previously-seen points y, these will all give $K(x, y) = -1/2$. In particular, this means that $F(x)$ is independent of the target function c. Finally, since X' has size $2^{n/2}$ and d is only polynomial in n, we have by simply counting the number of possible partitions of $F(X')$ by halfspaces that with high probability $F(P)$ will not even be weakly separable for a random function c over X'. Specifically, for any given halfspace, the probability over choice of c that it has error less than $1/2 - \epsilon$ is exponentially small in $|X'|$ (by Hoeffding bounds), which is doubly-exponentially small in n, whereas there are "only" $2^{O(dn)}$ possible partitions by halfspaces.   □

The above construction corresponds to a scenario in which "real data" (the points in X') are so sparse and special that an algorithm without access to D is not able to construct anything that looks even remotely like a real data point by itself (e.g., examples are pixel images of outdoor scenes and yet our poor learning algorithm has no knowledge of vision and can only construct white noise). Furthermore, it relies on a kernel that only does something interesting on the real data (giving nothing useful for $x \notin X'$). It is conceivable that positive results independent of the distribution D can be achieved for standard, natural kernels.

5   Conclusions and Open Problems

This survey has examined ways in which random projection (of various forms) can provide algorithms for, and insight into, problems in machine learning. For example, if a learning problem is separable by a large margin γ, then a random projection to a space of dimension $O(\frac{1}{\gamma^2} \log \frac{1}{\epsilon\delta})$ will with high probability approximately preserve separability, so we can think of the problem as really an $O(1/\gamma^2)$-dimensional problem after all. In addition, we saw that just picking a *random* separator (which can be thought of as projecting to a random 1-dimensional space) has a reasonable chance of producing a weak hypothesis.

We also saw how given black-box access to a kernel function K and a distribution D (i.e., unlabeled examples) we can use K and D together to construct a new low-dimensional feature space in which to place the data that approximately preserves the desired properties of the kernel. Thus, through this mapping, we can think of a kernel as in some sense providing a distribution-dependent feature space. One interesting aspect of the simplest method considered, namely choosing $x_1, \ldots, x_d$ from D and then using the mapping $x \mapsto (K(x, x_1), \ldots, K(x, x_d))$, is that it can be applied to any generic "similarity" function $K(x, y)$, even those that are not necessarily legal kernels and do not necessarily have the same interpretation as computing a dot-product in some implicit ϕ-space. Recent results of [BB06] extend some of these guarantees to this more general setting.

One concrete open question is whether, for natural standard kernel functions, one can produce mappings $F : X \to R^d$ in an oblivious manner, without using examples from the data distribution. The Johnson-Lindenstrauss lemma tells us that such mappings exist, but the goal is to produce them without explicitly computing the ϕ-function. Barring that, perhaps one can at least reduce the unlabeled sample-complexity of our approach. On the practical side, it would be interesting to further explore the alternatives that these (or other) mappings provide to widely used algorithms such as SVM and Kernel Perceptron.

Acknowledgements

Much of this was based on joint work with Maria-Florina (Nina) Balcan and Santosh Vempala. Thanks also to the referees for their helpful comments. This work was supported in part by NSF grants CCR-0105488, NSF-ITR CCR-0122581, and NSF-ITR IIS-0312814.

References

[Ach03]   D. Achlioptas. Database-friendly random projections. *Journal of Computer and System Sciences*, 66(4):671–687, 2003.

[AV99]   R. I. Arriaga and S. Vempala. An algorithmic theory of learning, robust concepts and random projection. In *Proceedings of the 40th Annual IEEE Symposium on Foundation of Computer Science*, pages 616–623, 1999.

[BB05]   M-F. Balcan and A. Blum. A PAC-style model for learning from labeled
           and unlabeled data. In *Proceedings of the 18th Annual Conference on
           Computational Learning Theory (COLT)*, pages 111–126, 2005.

[BB06]   M-F. Balcan and A. Blum. On a theory of kernels as similarity functions.
           Mansucript, 2006.

[BBV04]   M.F. Balcan, A. Blum, and S. Vempala. Kernels as features: On kernels,
           margins, and low-dimensional mappings. In *15th International Conference
           on Algorithmic Learning Theory (ALT '04)*, pages 194–205, 2004. An ex-
           tended version is available at http://www.cs.cmu.edu/~avrim/Papers/.

[BGV92]   B. E. Boser, I. M. Guyon, and V. N. Vapnik. A training algorithm for
           optimal margin classifiers. In *Proceedings of the Fifth Annual Workshop
           on Computational Learning Theory*, 1992.

[BIG74]   A. Ben-Israel and T.N.E. Greville. *Generalized Inverses: Theory and
           Applications*. Wiley, New York, 1974.

[Blo62]   H.D. Block. The perceptron: A model for brain functioning. *Reviews
           of Modern Physics*, 34:123–135, 1962. Reprinted in *Neurocomputing*,
           Anderson and Rosenfeld.

[BST99]   P. Bartlett and J. Shawe-Taylor. Generalization performance of support
           vector machines and other pattern classifiers. In *Advances in Kernel
           Methods: Support Vector Learning*. MIT Press, 1999.

[CV95]   C. Cortes and V. Vapnik. Support-vector networks. *Machine Learning*,
           20(3):273 – 297, 1995.

[Das00]   S. Dasgupta. Experiments with random projection. In *Proceedings of the
           16th Conference on Uncertainty in Artificial Intelligence (UAI)*, pages
           143–151, 2000.

[DG02]   S. Dasgupta and A. Gupta. An elementary proof of the Johnson-Lind-
           enstrauss Lemma. *Random Structures & Algorithms*, 22(1):60–65, 2002.

[EK00]   Y. Rabani E. Kushilevitz, R. Ostrovsky. Efficient search for approxi-
           mate nearest neighbor in high dimensional spaces. *SIAM J. Computing*,
           30(2):457–474, 2000.

[FM03]   D. Fradkin and D. Madigan. Experiments with random projections for
           machine learning. In *KDD '03: Proceedings of the ninth ACM SIGKDD
           international conference on Knowledge discovery and data mining*, pages
           517–522, 2003.

[FS97]   Y. Freund and R. Schapire. A decision-theoretic generalization of on-
           line learning and an application to boosting. *Journal of Computer and
           System Sciences*, 55(1):119–139, 1997.

[FS99]   Y. Freund and R.E. Schapire. Large margin classification using the Per-
           ceptron algorithm. *Machine Learning*, 37(3):277–296, 1999.

[GBN05]   N. Goal, G. Bebis, and A. Nefian. Face recognition experiments with
           random projection. In *Proceedings SPIE Vol. 5779*, pages 426–437, 2005.

[GW95]   M.X. Goemans and D.P. Williamson. Improved approximation algo-
           rithms for maximum cut and satisfiability problems using semidefinite
           programming. *Journal of the ACM*, pages 1115–1145, 1995.

[IM98]   P. Indyk and R. Motwani. Approximate nearest neighbors: towards re-
           moving the curse of dimensionality. In *Proceedings of the 30th Annual
           ACM Symposium on Theory of Computing*, pages 604–613, 1998.

[JL84]   W. B. Johnson and J. Lindenstrauss. Extensions of Lipschitz mappings
           into a Hilbert space. In *Conference in Modern Analysis and Probability*,
           pages 189–206, 1984.

[Lit89]   Nick Littlestone. From on-line to batch learning. In *COLT '89: Proceedings of the 2nd Annual Workshop on Computational Learning Theory*, pages 269–284, 1989.

[MMR+01]   K. R. Muller, S. Mika, G. Ratsch, K. Tsuda, and B. Scholkopf. An introduction to kernel-based learning algorithms. *IEEE Transactions on Neural Networks*, 12:181–201, 2001.

[MP69]   M. Minsky and S. Papert. *Perceptrons: An Introduction to Computational Geometry*. The MIT Press, 1969.

[Nov62]   A.B.J. Novikoff. On convergence proofs on perceptrons. In *Proceedings of the Symposium on the Mathematical Theory of Automata, Vol. XII*, pages 615–622, 1962.

[Sch90]   R. E. Schapire. The strength of weak learnability. *Machine Learning*, 5(2):197–227, 1990.

[Sch00]   L. Schulman. Clustering for edge-cost minimization. In *Proceedings of the 32nd Annual ACM Symposium on Theory of Computing*, pages 547–555, 2000.

[STBWA98]   J. Shawe-Taylor, P.L. Bartlett, R.C. Williamson, and M. Anthony. Structural risk minimization over data-dependent hierarchies. *IEEE Trans. on Information Theory*, 44(5):1926–1940, 1998.

[Vap98]   V. N. Vapnik. *Statistical Learning Theory*. John Wiley and Sons Inc., New York, 1998.

[Vem98]   S. Vempala. Random projection: A new approach to VLSI layout. In *Proceedings of the 39th Annual IEEE Symposium on Foundation of Computer Science*, pages 389–395, 1998.

[Vem04]   S. Vempala. *The Random Projection Method*. American Mathematical Society, DIMACS: Series in Discrete Mathematics and Theoretical Computer Science, 2004.

Some Aspects of Latent Structure Analysis

D.M. Titterington

University of Glasgow, Glasgow, Scotland, UK
mike@stats.gla.ac.uk

Abstract. Latent structure models involve real, potentially observable variables and latent, unobservable variables. The framework includes various particular types of model, such as factor analysis, latent class analysis, latent trait analysis, latent profile models, mixtures of factor analysers, state-space models and others. The simplest scenario, of a single discrete latent variable, includes finite mixture models, hidden Markov chain models and hidden Markov random field models. The paper gives a brief tutorial of the application of maximum likelihood and Bayesian approaches to the estimation of parameters within these models, emphasising especially the fact that computational complexity varies greatly among the different scenarios. In the case of a single discrete latent variable, the issue of assessing its cardinality is discussed. Techniques such as the EM algorithm, Markov chain Monte Carlo methods and variational approximations are mentioned.

1 Latent-Variable Fundamentals

We begin by establishing notation. Let y denote observable data for an experimental unit, let z denote missing or otherwise unobservable data and let $x = (y, z)$ denote the corresponding complete data. Then probability functions (densities or mass functions according as the variables are discrete or continuous) are indicated as follows: $f(y, z)$ for y and z jointly, $g(y|z)$ for y conditional on z, and $f(y)$ and $h(z)$ as marginals for y and z respectively. These functions satisfy the relationships

$$f(y) = \int f(y, z) dz$$
$$= \int g(y|z) h(z) dz,$$

where the integration is replaced by summation if z is discrete. Also of interest might be the other conditional probability function, $h(z|y)$, which can be expressed as

$$h(z|y) = f(y, z)/f(y) \propto g(y|z) h(z),$$

which is essentially an expression of Bayes' Theorem.

In latent-structure contexts, z represents latent variables, introduced to create flexible models, rather than 'real' items, although often real, physical interpretations are surmised for the latent variables, in the case of factor analysis,

C. Saunders et al. (Eds.): SLSFS 2005, LNCS 3940, pp. 69–83, 2006.

for example. If $y = (y_1, \ldots, y_p)$ is p-dimensional, corresponding to p observable characteristics, then Bartholomew [1] proposes that

$$g(y|z) = \prod_i g_i(y_i|z),$$

which ensures that it is the dependence of the y_i on the latent variables z that accounts fully for the mutual dependence among the y_i.

2   Simple Particular Cases

There are a number of simple particular cases, depending on the natures of the observed variables y and the latent variables z:

- the case of y continuous and z continuous corresponds to *factor analysis*;
- the case of y continuous and z discrete corresponds to *latent profile analysis* [2] or *cooperative vector quantisation* [3];
- the case of y discrete and z continuous corresponds to *latent trait analysis* [4] or *density networks* [5];
- the case of y discrete and z discrete corresponds to *latent class analysis* [6] or *naive Bayesian networks* [7].

It is noteworthy that, in three of the cases, two nomenclatures are given for the same structure, one from the statistics literature and one more prevalent in the machine learning/computer science literature; this emphasises the fact that these models are of common interest to these two research communities.

A number of general remarks can be made: in all but factor analysis z is usually univariate, although review and development of the case of multivariate z is provided by Dunmur and Titterington [8]; if however z is multivariate then its components are usually assumed to be independent; continuous variables are usually assumed to be Gaussian, especially within z; and discrete variables are usually categorical or binary, rather than numerical or ordinal. Other variations include *mixtures of factor-analysers*, in which y contains continuous variables and z includes some continuous variables and at least one categorical variable [9] [10].

We now present the above four special cases in the form of statistical models.

1. *Factor analysis:*
$$y = Wz + e,$$
where $z \sim N(0, I)$ and $e \sim N(0, \Lambda)$ with z and e independent and Λ diagonal. The dimensionality of z is normally less than p, the dimensionality of y.

2. *Latent profile analysis:*
$$y = WZ + e,$$
where Z is a matrix of indicators for multinomially generated z and e is as in factor analysis.

3. *Latent trait analysis:*

$$g(y_{is} = 1|z) \propto \exp(w_{0is} + w_{is}^\top z),$$

where $\{y_{is}\}$ are elements of an indicator vector representing the ith component of y. There are obvious constraints, in that we must have $\sum_s g(y_{is} = 1|z) = 1$ for each i. Also, for identifiability, constraints such as $w_{0i1} = 0, w_{i1} = 0$, for each i, must be imposed. Thus the conditional distributions associated with y given z correspond to linear logistic regression models.

4. *Latent class analysis:*
The conditional probabilities $g(y_{is} = 1|z = u)$ are multinomial probabilities that sum to 1 over s, for each i, and $h(z = u)$ are multinomial probabilities that sum to 1 over u.

3   Issues of Interest

At a general level there are two main issues, listed here in arguably the reverse of the correct order of fundamental importance!

- Estimation of the chosen model, i.e. of relevant parameters, according to some paradigm; we shall describe likelihood-based and Bayesian approaches. By 'parameters' we mean items such as W and Λ in the factor analysis model, and so on.
- Estimation or selection of the model structure itself, and in particular the appropriate level of complexity, in some sense, of z: the more complex the structure of z is, the more 'flexible' is the latent-structure model for y.

Inference will be based on representative data, possibly supplemented with 'prior' information. The data will consist of a number of realisations of the observables:

$$D_y = \{y^{(n)}, n = 1, \ldots, N\}.$$

Thus N represents the number of experimental units in the dataset and often corresponds to 'sample size' in statistical terminology. The fact that the data are 'incomplete', with the latent variables $D_z = \{z^{(n)}, n = 1, \ldots, N\}$ being unobservable, complicates matters, but so also might other aspects of the model structure, as we shall see.

Suppose that $D_x = \{(y^{(n)}, z^{(n)}), n = 1, \ldots, N\}$ denotes the *complete data*, and that the total set of parameters is

$$\theta = (\phi, \eta),$$

where ϕ and η denote parameters within the models for $y|z$ and z respectively. (In this respect the factor analysis model is rather special, in that $\phi = (W, \Lambda)$ and there is no unknown η.) Then likelihood and Bayesian inference for θ should be based on the observed-data likelihood, defined by the marginal density associated with the observed data but regarded as a function of θ,

$$f(D_y|\theta) = \sum_{D_z} f(D_x|\theta), \tag{1}$$

where here we are assuming that the variables within z are discrete and, in expanded form,

$$f(D_x|\theta) = g(D_y|D_z, \phi)h(D_z|\eta).$$

Note that Bayesian inference about θ should be based on the posterior density

$$p(\theta|D_y) \propto f(D_y|\theta)\, p(\theta),$$

where $p(\theta)$ is a prior density for θ.

Were D_z not latent but known, then the basis for inference would be the usually much simpler $f(D_x|\theta)$, in the case of likelihood inference, or the corresponding $p(\theta|D_x)$ if the Bayesian approach is being adopted. The marginalisation operation depicted in (1) typically creates quite a complicated object.

4   The Case of a Single Categorical Latent Variable

Suppose that each hidden $z^{(n)}$ is categorical, with K categories, denoted by $\{1, \ldots, K\}$, and that the conditional density of $y^{(n)}$, given that $z^{(n)} = k$, is the *component density* $g_k(y^{(n)}|\phi_k)$. Often Gaussian component densities are used, in which case ϕ_k contains the mean vector and covariance matrix of the kth component density.

As we have just seen, the function of key interest, as the observed-data likelihood, is the joint density for D_y. The complexity of this density will be dictated by the pattern of dependence, marginally, among the $y^{(n)}$'s. We shall assume that the different $y^{(n)}$'s are conditionally independent, given the $z^{(n)}$'s and that $y^{(n)}$ depends only on $z^{(n)}$ and not on any other part of D_z, so that

$$g(D_y|D_z, \phi) = \prod_n g(y^{(n)}|z^{(n)}, \phi);$$

we shall consider different possibilities for the dependence structure among the $z^{(n)}$'s.

1. *Case 1: $z^{(n)}$'s independent.* If the $z^{(n)}$'s are independent then so, marginally, are the $y^{(n)}$'s:

$$f(D_y|\theta) = \sum_{D_z} f(D_y, D_z|\theta)$$

$$= \sum_{D_z} g(D_y|D_z, \phi)h(D_z|\eta)$$

$$= \sum_{D_z} \{\prod_{n=1}^{N} g(y^{(n)}|z^{(n)}, \phi) \prod_{n=1}^{N} h(z^{(n)}|\eta)\}$$

$$= \prod_{n=1}^{N} \{ \sum_{k=1}^{K} g_k(y^{(n)}|\phi_k)\eta_k \},$$

where $\eta_k = \text{Prob}(z^{(n)} = k)$. In this case therefore the observed data constitute a sample of size N from a *mixture distribution*, which might otherwise
be called a *hidden multinomial*. The mixture density is

$$f(y|\theta) = \sum_{k=1}^{K} g_k(y|\phi_k) \, \eta_k,$$

and the $\{\eta_k\}$ are called the *mixing weights*.

2. *Case 2: $z^{(n)}$'s following a Markov chain.* In this case the $y^{(n)}$'s correspond
   to a *hidden Markov (chain) model*, much applied in economics and speech-
   modelling contexts. (The version with continuous (D_y, D_z) corresponds to
   state-space dynamic models.) In this case the computation of the observed-
   data likelihood $f(D_y|\theta)$ is more complicated, essentially because the sum-
   mation operation cannot be dealt with so simply, but in principle $f(D_y|\theta)$
   can be computed by a pass through the data.
3. *Case 3: $z^{(n)}$'s following a Markov random field.* In this case the index set is
   typically two-dimensional, corresponding to a lattice of N grid points. The
   $y^{(n)}$'s correspond to a noisy/hidden Markov random field model popular in
   the statistical analysis of pixellated images [11]: here D_z represents the true
   scene and D_y a noise-corrupted but observable version thereof. This scenario
   includes simple versions of Boltzmann machines. The computation of the
   observed-data likelihood $f(D_y|\theta)$ is typically not a practical proposition.

The next two sections deal with inference paradigms, and we shall see that in
each case the level of difficulty escalates as our attention moves from Case 1 to
Case 2 to Case 3, as has just been mentioned in the context of the calculation
of $f(D_y|\theta)$.

5   Maximum Likelihood Estimation

5.1   The EM Algorithm

The EM algorithm [12] is an iterative algorithm that aims to converge to maximum likelihood estimates in contexts involving incomplete data. From an initial
approximation $\theta^{(0)}$, the algorithm generates a sequence $\{\theta^{(r)}\}$ using the following iterative double-step.

E-Step. Evaluate

$$Q(\theta) = E\{\log f(D_x|\theta)|D_y, \theta^{(r)}\}.$$

M-Step. Calculate

$$\theta^{(r+1)} = \arg\max_\theta Q(\theta).$$

Typically, $f(D_y|\theta^{(r+1)}) \geq f(D_y|\theta^{(r)})$, so that the likelihood of interest increases at each stage and, although there are exceptions to the rule, the algorithm converges to at least a local, if not a global, maximum of the likelihood. In summary, the E-Step calculates a (conditional) expectation of the corresponding complete-data loglikelihood function and the M-step maximises that function. The convenience of the algorithm depends on the ease with which the E-Step and M-Step can be carried out, and we discuss this briefly in the context of the three cases identified in the previous section.

So far as the M-Step is concerned, the level of difficulty is the same as that which applies in the corresponding complete-data context. In most mixture problems and most hidden Markov chain contexts this is easy, but in the context of hidden Markov random fields aspects of the M-Step are very difficult, as we shall shortly illustrate. The E-Step for these models amounts to the calculation of expectations of the components of the indicator variables D_z; these expectations are therefore probabilities of the various possible configurations for the latent states, given the observed data. Difficulties in the E-Step are usually caused by intractability of the distribution of $D_x|D_y, \theta^{(r)}$ or equivalently of $D_z|D_y, \theta^{(r)}$. As a result of the independence properties with mixture data, the distribution of $D_z|D_y, \theta^{(r)}$ becomes the product model corresponding to the marginal distributions for $z^{(n)}|y^{(n)}, \theta^{(r)}$, for each n, and the E-Step in this case is straightforward. For the hidden Markov chain case, the dependence among the $z^{(n)}$'s does complicate matters, but the dependence is Markovian and 'one-dimensional', and this leads to the E-Step being computable by a single forwards and then backwards pass through the data [13]. This case is therefore less trivial than for mixtures but is not a serious problem. In the hidden Markov random field case, however, the E-Step is dramatically more difficult; the dependence among the $z^{(n)}$'s is still Markovian, but is 'two-dimensional', and there is no simple analogue of the forwards-backwards algorithm. We illustrate the difficulties in both steps with the simplest example of a hidden Markov random field.

Example. Hidden Ising model.
Suppose the index set for D_z is a two-dimensional lattice and that

$$h(D_z) = h(D_z|\eta) = \{G(\eta)\}^{-1}\exp\{\eta \sum_{s \sim t} z^{(s)} z^{(t)}\},$$

where each $z^{(n)} \in \{-1, +1\}$, so that each hidden variable is binary, η is a scalar parameter, usually positive so as to reflect local spatial association, and the summation is over (s, t) combinations of locations that are immediate vertical or horizontal neighbours of each other on the lattice; this constitues the Ising model of statistical physics. In this model the normalising constant $G(\eta)$ is not computable. This leads to there being no analytical form for the E-step and no easy M-Step for η. For example, as mentioned earlier, the degree of difficulty of the M-Step for η is the same as that of complete-data maximum likelihood, and for the latter one would have to maximise $h(D_z|\eta)$ with respect to η, which is stymied by the complexity of $G(\eta)$.

What can be done if the EM algorithm becomes impracticable? A number of possible approaches exist.

- With the Law of Large Numbers in mind, replace the E-Step by an appropriate sample mean calculated from realisations of the conditional distribution of D_z given D_y and $\theta^{(r)}$. However, in the context of the hidden Ising model, simulation from the relevant conditional distribution is not straightforward and itself requires an iterative algorithm of the Markov chain Monte Carlo type.
- Replace the complicated conditional distribution in the E-Step by a deterministic, simpler approximation, e.g. a variational approximation. We give more details of this in the next subsection, concentrating on it because of its prominence in the recent computer science literature.
- Use variational (or other) approximations in the M-Step.
- Other suggestions exist including the consideration of methods other than EM. For example, Younes [14] developed a gradient-based stochastic approximation method and Geyer and Thompson [15] used Monte Carlo methods to approximate the intractable normalisation constant and thereby attack the likelihood function directly. In [16] a number of more ad hoc methods are suggested for the image-analysis context, based on iterative restoration of the true scene, perhaps using Besag's [17] Iterative Conditional Modes algorithm, alternated with parameter estimation with the help of Besag's *pseudolikelihood* [18] for the parameter η. The simplest form of the pseudolikelihood is the product of the full conditional densities for the individual $z^{(n)}$'s, given the rest of D_z. It is much easier to handle than the original $h(D_z|\eta)$ because there is no intractable normalisation constant, and yet the maximiser of the pseudolikelihood is generally a consistent estimator of the true η.

5.2   Variational Approximations

Suppose that $h(D_z)$ is a complicated multivariate distribution, and that $q(D_z)$ is a proposed tractable approximation to $h(D_z)$ with a specified structure. Then one way of defining an optimal q of that structure is to minimise an appropriate measure of distance between q and h, such as the Kullback-Leibler (KL) divergence,

$$\mathrm{KL}(q, h) = \sum_{D_z} q(D_z) \log \{q(D_z)/h(D_z)\}.$$

Often the form of q is determined by the solution of this variational optimisation exercise, as are the values of (variational) hyperparameters that q contains. The simplest model for q would be an independence model, i.e.

$$q(D_z) = \prod_n q_n(z^{(n)}),$$

which leads to *mean field approximations*. Furthermore, variational approximations to the conditional distribution of D_z given D_y lead to lower bounds on the observed-data loglikelihood. To see this note that

$$\log f(D_y|\theta) = \log \left\{ \sum_{D_z} f(D_y, D_z|\theta) \right\}$$

$$\geq \sum_{D_z} q(D_z) \log \{ f(D_y, D_z|\theta)/q(D_z) \},$$

by Jensen's inequality. Typically, q is chosen to have a structure such that the summation in the lower bound is easily achieved; the choice of a fully factorised q is certainly advantageous in this respect, but it represents a simplifying approximation whose consequences should be investigated. It is straightforward to show that the q, of any prescribed structure, that maximises the above lower bound for $\log f(D_y|\theta)$, minimises the KL divergence between q and the conditional distribution for D_z given D_y and θ. An EM-like algorithm can be evolved in which q and θ are successively updated in the equivalents of the E-Step and M-Step respectively. Convergence of the resulting sequence of iterates for θ, to a local maximum of the lower-bound surface, can be proved, but there is comparatively little theory about the relationship of such a maximum to the maximiser of the observed-data likelihood itself, although some progress is reported in [19]. (If one can show that the maximiser of the lower-bound function tends to the maximiser of the likelihood, asymptotically, then one can claim that the lower-bound maximiser inherits the maximum likelihood estimator's property of being consistent for the true value of θ.)

The practicality of variational approximations relies on the computability of the lower-bound function, and much of the relevant literature is restricted to the case of a fully factorised q_{D_z}. However, more-refined approximations can be developed in some contexts [20] [21]. It would also be of obvious value to obtain corresponding upper bounds for $\log f(D_y|\theta)$, but they are much harder to come by and a general method for deriving them is as yet elusive.

Wainwright and Jordan [22] take a more general convex analysis approach to defining variational approximations, although operational versions of the method usually amount to optimising a KL divergence.

For an application of these ideas to latent profile analysis see [8], and for a tutorial introduction to variational approximations see [23]. In earlier work, Zhang [24] [25] used mean-field-type approximations within the EM-algorithm in contexts such as image restoration based on underlying Markov random field models.

6   The Bayesian Approach

6.1   Introduction

As already stated, Bayesian inference for the parameters in a model is based on

$$p(\theta|D_y) \propto f(D_y|\theta) \, p(\theta) = \left\{ \sum_{D_z} f(D_x|\theta) \right\} p(\theta),$$

where $p(\theta)$ is a prior density for θ. As with maximum likelihood, Bayesian analysis is often vastly easier if D_z is not missing, in which case we use

$$p(\theta|D_x) \propto f(D_x|\theta)\, p(\theta).$$

In many familiar cases, $f(D_x|\theta)$ corresponds to an exponential family model and then there exists a family of neat closed-form *conjugate* priors for θ; the posterior density $p(\theta|D_x)$ then belongs to the same conjugate family, with hyperparameters that are easily written down; see for example Section 3.3 of [26]. This convenient pattern disappears if there are missing data, in this represented by D_z. Inevitably non-exact methods are then needed, and most common approaches can be categorised either as asymptotically exact but potentially unwieldy simulation methods, usually Markov chain Monte Carlo (MCMC) methods, or as non-exact but less unwieldy deterministic approximations. The latter include Laplace approximations [27] and variational Bayes approximations; we shall concentrate on the latter, again because of its high profile in recent machine-learning literature.

Both these approaches aim to approximate

$$p(\theta, D_z|D_y) \propto f(D_x|\theta)\, p(\theta),$$

the joint posterior density of all unknown items, including the latent variables as well as the parameters. Marginalisation then provides an approximation to $p(\theta|D_y)$.

6.2   The MCMC Simulation Approach

This method aims to generate a set of simulated realisations from $p(\theta, D_z|D_y)$. The resulting set of realisations of θ then form a sample from $p(\theta|D_y)$, and, for example, the posterior mean of θ can be approximated by the empirical average of the realisations of θ.

The now-standard approach to this is to generate a sequence of values of the variables of interest from a Markov chain for which the equilibrium distribution is $p(\theta, D_z|D_y)$. Once the equilibrium state has been reached, a realisation from the required distirubtion has been generated. There are various general recipes for formulating such a Markov chain, one of the simplest being *Gibbs sampling*. This involves recursively sampling from the appropriate set of full conditional distributions of the unknown items. A 'block' version of this for our problem would involve iteratively simulating from $p(\theta|D_z, D_y)$ and $p(D_z|\theta, D_y)$. This is typically easy for mixtures [28] and hidden Markov chains [29] but is problematic with hidden Markov random fields [30]. As with maximum likelihood, the intractability of the normalising constant is the major source of the difficulty.

The development and application of MCMC methods in Bayesian statistics has caught on spectacularly during the quarter-century since the publication of papers such as [31]; see for example [32] [33]. However, important issues remain that are the subject of much current work, including the monitoring of convergence, difficulties with large-scale problems, the invention of new samplers, and the search for perfect samplers for which convergence at a specified stage can be guaranteed. A variety of approximate MCMC approaches are described and compared in [34].

6.3   Variational Bayes Approximations

Suppose that $q(\theta, D_z)$ defines an approximation to $p(\theta, D_z | D_y)$ and suppose we *propose* that q take the factorised form

$$q(\theta, D_z) = q_\theta(\theta) q_{D_z}(D_z).$$

Then the factors are chosen to minimise

$$\mathrm{KL}(q, p) = \int_\theta \sum_{D_z} q \, \log \, (q/p),$$

in which p is an abbreviation for $p(\theta, D_z | D_y)$. The resulting $q_\theta(\theta)$ is regarded as the approximation to $p(\theta | D_y)$.

The form of q_θ is often the same as that which would result in the complete-data case: if a prior $p(\theta)$ is chosen that is conjugate for the Bayesian analysis of the complete data, D_x, then q_θ also takes that convenient conjugate form but calculation of the (hyper)parameters within q_θ requires the solution of nonlinear equations.

Once again, the analysis for mixtures and hidden Markov chains is comparatively 'easy', but the case of hidden Markov random fields is 'hard'; see for example [35] [36] [37] [38]. As in the case of likelihood-based variational approximations, there is not much work on the theoretical properties of the method. However, in a number of scenarios, including Gaussian mixture distributions, Wang and Titterington have shown that the variational posterior mean is consistent, see for example [39], but they have also shown that the variational posterior variances can be unrealistically small, see for example [40].

For a review of variational and other approaches to Bayesian analysis in models of this general type see [41].

7   A Brief Discussion of Model Selection

7.1   Non-bayesian Approaches

We continue to concentrate on the case of a single categorical latent variable, and especially on mixture models. The model-selection issue of interest will be the determination of an appropriate number K of components to be included in the model. General non-Bayesian approaches include the following:

- selection of a parsimonious model, i.e. the minimum plausible K, by hypothesis-testing;
- optimisation of criteria such as Akaike's AIC [42] and Schwarz's BIC [43].

However, it is well known that standard likelihood-ratio theory for nested models, based on the use of chi-squared distributions for testing hypotheses, breaks down with mixture models. Various theoretical and practical directions have been followed for trying to overcome this; among the latter is McLachlan's [44] use of bootstrap tests for mixtures.

The criteria AIC and BIC impose penalties on the maximised likelihood to penalise highly-parameterised models, so it is plausible that these instruments should select a suitable value of K. This is by no means guaranteed, especially for AIC, but, asymptotically at least, there are results showing that BIC selects the right number of mixture components in at least some cases [45]. It should be pointed out that AIC was developed with scenarios in mind in which, unlike in the case of mixtures, there is the same sort of 'regularity' as is required for standard likelihood-ratio theory.

7.2   Bayesian Approaches

We consider the following three approaches:

- model comparison using Bayes factors;

- use of Bayesian-based selection criteria (DIC);

- generation of a posterior distribution for the cardinality of the latent space.

Approach 1: Bayes factors.
Suppose θ_k represents the parameters present in Model M_k, and that $\{p(M_k)\}$ are a set of prior probabilities over the set of possible models. Then the ratio of posterior probabilities for two competing models M_k and $M_{k'}$ is given by

$$\frac{p(M_k|D_y)}{p(M_{k'}|D_y)} = \frac{f(D_y|M_k)}{f(D_y|M_{k'})} \times \frac{p(M_k)}{p(M_{k'})},$$

where the first ratio on the right-hand side is the *Bayes factor*, and

$$f(D_y|M_k) = \int f(D_y|\theta_k)d\theta_k,$$

in which θ_k are the parameters corresponding to model M_k. A 'first-choice' model would be one for which the posterior probability $p(M_k|D_y)$ is maximum. Clearly the calculation of the Bayes factor is complicated in incomplete-data scenarios. For an authoritative account of Bayes factors see [46].

Approach 2: The Deviance Information Criterion.
This can be described as a Bayesian version of AIC. First we define 'Deviance' by

$$\Delta(\theta) = -2 \log \{f(D_y|\theta)\},$$

and define

$$\overline{\Delta(\theta)} = E_{p(\theta|D_y)}\Delta(\theta)$$
$$\bar{\theta} = E_{p(\theta|D_y)}\theta$$
$$p_\Delta = \overline{\Delta(\theta)} - \Delta(\bar{\theta}).$$

Then DIC is defined by

$$\text{DIC} = \overline{\Delta(\theta)} + p_\Delta,$$

clearly to be minimised if model selection is the aim. The method was introduced and implemented in a variety of contexts, mainly involving complete data from generalised linear models, in [47]. A number of somewhat ad hoc adaptations for incomplete-data contexts were proposed and compared, in particular in mixture models, in [48], and C.A. McGrory's Thesis [38] will report on the application of variational approximations in the context of DIC, using mixture models, hidden Markov chains and hidden Markov random fields as testbeds.

Approach 3 : Generation of $\{p(k|D_y)\}$.
As remarked when we were discussing Bayes factors, 'exact' computation of $\{p(k|D_y)\}$ is very difficult in incomplete-data contexts, so instead attempts have been made to assess it using MCMC. Two strands have been developed.

1. *Reversible Jump MCMC* [49]. As in 'ordinary' MCMC, a Markov chain is generated that is designed to have, as equilibrium distribution, the posterior distribution of all unknown quantities, *including* k. Therefore, since the value of k can change during the procedure, there is the need to jump (reversibly) between parameter spaces of different dimensions, and this creates special problems. (The reversibility property is required in order to guarantee convergence of the Markov chain Monte Carlo procedure to the desired equilibrium distribution.) Of the problems covered in the present paper, mixtures are dealt with in [50], hidden Markov chains in [51], and some comparatively small-scale spatial problems in [52].
2. *Birth and Death MCMC* [53]. This can be thought of as a 'continuous-time' alternative to reversible jump MCMC. The key feature is the modelling of the mixture components as a marked birth-and-death process, with components being added (birth) of being discarded (death) according to a random process, with 'marks' represented by the parameters of the component distributions.

Cappé et al. [54] showed that the two approaches could be linked and generalised.

Acknowledgement

This work was supported by the IST Programme of the European Community, under the PASCAL Network of Excellence, IST-2002-506778. This publication only reflects the author's views. The paper was prepared for a PASCAL workshop in Bohinj, Slovenia, in February, 2005.

References

1. Bartholomew, D.J.: The foundations of factor analysis. Biometrika **71** (1984) 221–232.
2. Gibson, W.A.: Three multivariate models: factor analysis, latent structure analysis and latent profile analysis. Psychometrika **24** (1959) 229–252.
3. Ghahramani, Z.: Factorial learning and the EM algorithm. In Advances in Neural Information Processing Systems **7** (eds. G. Tesauro, D.S. Touretzky and T.K. Leen). (1996) MIT Press, Cambridge MA.

4. Bartholomew, D.J.: Latent Variable Models and Factor Analysis. (1987) Griffin, London.
5. MacKay, D.J.C.: Bayesian neural networks and density networks. Instr. Meth. Phys. Res. A **354** (1995) 73–80.
6. Hagenaars, J.A.: Categorical Longitudinal Data. (1990) Sage, London.
7. Neal, R.M.: Probabilistic inference using Markov chain Monte Carlo methods. (1993) Tech. Report CRG-TR-93-1, Dept. Comp. Sci., Univ. Toronto.
8. Dunmur, A.P., Titterington, D.M.: Analysis of latent structure models with multi-dimensional latent variables. In Statistics and Neural Networks: Recent Advances at the Interface (eds. J.W. Kay and D.M. Titterington) (1999) 165–194. Oxford: Oxford University Press.
9. Ghahramani, Z., Beal, M.: Variational inference for Bayesian mixtures of factor analyzers. In Advances in Neural Information Processing, Vol.12 (eds., S.A. Solla, T.K. Leen and K.-R. Müller) (2000) 449–455. MIT Press, Cambridge, MA.
10. Fokoué, E., Titterington, D.M.: Mixtures of factor analysers: Bayesian estimation and inference by stochastic simulation. Machine Learning **50** (2003) 73–94.
11. Geman, S., Geman, D.: Stochastic relaxation, Gibbs distributions, and the Bayesian restoration of images. IEEE Trans. Patt. Anal. Mach. Intell. **6** (1984) 721–741.
12. Dempster, A.P., Laird, N.M., Rubin, D.B.: Maximum likelihood from incomplete data via the EM algorithm (with discussion). J. R. Statist. Soc. B **39** (1977) 1–38.
13. Rabiner, L.: A tutorial on hidden Markov models and selected applications in speech recognition. Proc. IEEE **77** (1989) 257–285.
14. Younes, L.: Parameter estimation for imperfectly observed Gibbsian fields. Prob. Theory Rel. Fields **82** (1989) 625–645.
15. Geyer, C.J., Thompson, E.A.: Constrained Monte Carlo maximum likelihood for dependent data (with discussion). J.R. Statist. Soc. B **54** (1992) 657–699.
16. Qian, W., Titterington, D.M.: Estimation of parameters in hidden Markov models. Phil. Trans. R. Soc. Lond. A **337** (1991) 407–428.
17. Besag, J.E.: On the statistical analysis of dirty pictures (with discussion). J.R. Statist. Soc. B **48** (1986) 259–302.
18. Besag, J.E.: Statistical analysis of non-lattice data. The Statistician **24** (1975) 179–195.
19. Hall, P., Humphreys, K., Titterington, D.M.: On the adequacy of variational lower bounds for likelihood-based inference in Markovian models with missing values. J. R. Statist. Soc. B **64** (2002) 549–564.
20. Bishop, C.M., Lawrence, N., Jaakkola, T.S., Jordan, M.I.: Approximating posterior distributions in belief networks using mixtures. In Advances in Neural Information Processing Systems, Vol. 10 (eds. M.I. Jordan, M.J. Kearns and S.A. Solla) (1998) 416–422. MIT Press, Cambridge, MA.
21. Humphreys, K., Titterington, D.M.: Improving the mean field approximation in belief networks using Bahadur's reparameterisation of the multivariate binary distribution. Neural Processing Lett. **12** (2000) 183–197.
22. Wainwright, M.J., Jordan, M.I.: Graphical models, exponential families, and variational approximations. Technical Report 649, (2003) Dept. Statistics, Univ. California, Berkeley.
23. Jordan, M.I., Gharamani, Z., Jaakkola, T.S., Saul, L.K.: An introduction to variational methods for graphical models. In Learning in Graphical Models (ed. M. Jordan) (1999) 105–162. MIT Press, Cambridge, MA.
24. Zhang, J.: The Mean Field Theory in EM procedures for Markov random fields. IEEE Trans. Signal Processing **40** (1992) 2570–2583.

25. Zhang, J.: The Mean Field Theory in EM procedures for blind Markov random field image restoration. IEEE Trans. Image Processing **2** (1993) 27–40.
26. Robert, C.P.: The Bayesian Choice, 2nd ed. (2001) Springer.
27. Tierney, L., Kadane, J.B.: Accurate approximations to posterior moments and marginal densities. J. Amer. Statist. Assoc. **81** (1986) 82–86.
28. Diebolt, J., Robert, C.P.: Estimation of finite mixture distributions through Bayesian sampling. J.R. Statist. Soc. B **56** (1994) 363–375.
29. Robert, C.P., Celeux, G., Diebolt, J.: Bayesian estimation of hidden Markov chains: a stochastic implementation. Statist. Prob. Lett. **16** (1993) 77–83.
30. Rydén, T., Titterington, D.M.: Computational Bayesian analysis of hidden Markov models. J. Comp. Graph. Statist. **7** (1998) 194–211.
31. Gelfand, A.E., Smith, A.F.M.: Sampling-based approaches to calculating marginal densities. J. Amer. Statist. Assoc. **85** (1990) 398–409.
32. Gilks, W.R., Richardson, S., Spiegelhalter, D.J. (eds.): Markov Chain Monte Carlo in Practice. Chapman and Hall.
33. Doucet, A., de Freitas, N., Gordon, N. (eds.): Sequential Monte Carlo Methods in Practice. Springer.
34. Murray, I., Ghahramani, Z.: Bayesian learning in undirected graphical models: approximate MCMC algorithms. In Proc. 20th Conf. Uncertainty in Artificial Intell. (eds. M. Chickering and J. Halperin) (2004) 577–584. AUAI Press.
35. Corduneanu, A., Bishop, C.M.: Variational Bayesian model selection for mixture distributions. In Proc. 8th Int. Conf. Artific. Intell. Statist. (eds. T. Richardson and T. Jaakkola) (2001) 27–34. Morgan Kaufmann, San Mateo, CA.
36. Ueda, N., Ghahramani, Z.: Bayesian model search for mixture models based on optimizing variational bounds. Neural Networks **15** (2003) 1223–1241.
37. MacKay, D.J.C.: Ensemble learning for hidden Markov models. (1997) Technical Report, Cavendish Lab., Univ. Cambridge.
38. McGrory, C.A.: Ph.D. Dissertation (2005) Dept. Statist., Univ. Glasgow.
39. Wang, B., Titterington, D.M.: Convergence properties of a general algorithm for calculating variational Bayesian estimates for a normal mixture model. Bayesian Analysis **1** (2006) to appear.
40. Wang, B., Titterington, D.M.: Variational Bayes estimation of mixing coefficients. In Proceedings of a Workshop on Statistical Learning. Lecture Notes in Artificial Intelligence Vol. 3635, (eds. J. Winkler, M. Niranjan and N. Lawrence) (2005) pp. 281–295. Springer.
41. Titterington, D.M.: Bayesian methods for neural networks and related models. Statist. Sci. **19** (2004) 128–139.
42. Akaike, H,: Information theory and an extension of the maximum likelihood principle. In Proc. 2nd Int. Symp. Info. Theory (eds. B.N. Petrov and F. Csaki) (1973) 267–281. Budapest: Akadémiai Kiadó.
43. Schwarz, G.: Estimating the dimension of a model. Ann. Statist. **6** (1978) 461–466.
44. McLachlan, G.J.: On bootstrapping the likelihood ratio test statistics for the number of components in a normal mixture. Appl. Statist. **36** (1987) 318–324.
45. Keribin, C.: Consistent estimation of the order of mixture models. Sankhya A **62** (2000) 49–66.
46. Kass, R.E., Raftery, A.: Bayes factors. J. Amer. Statist. Assoc. **90** (1995) 773–795.
47. Spiegelhalter, D.J., Best, N.G., Carlin, B.P., van der Linde, A.: Bayesian measures of complexity and fit (with discussion). J. R. Statist. Soc. B **64** (2002) 583–639.
48. Celeux, G., Forbes, F., Robert, C.P., Titterington, D.M.: Deviation information criteria for missing data models. (2005) Submitted.

49. Green, P.J.: Reversible jump Markov chain Monte Carlo computation and Bayesian model determination. Biometrika **82** (1995) 711–732.
50. Richardson, S., Green, P.J.: On Bayesian analysis of mixtures with an unknown number of components (with discussion). J. R. Statist. Soc. B **59** (1997) 731–792.
51. Robert, C.P., Rydén, T., Titterington, D.M.: Bayesian inference in hidden Markov models through the reversible jump Markov chain Monte Carlo method. J. R. Statist. Soc. B **62** (2000) 57–75.
52. Green, P.J., Richardson, S.: Hidden Markov models and disease mapping. J. Amer. Statist. Assoc. **97** (2002) 1055–1070.
53. Stephens, M.: Bayesian analysis of mixtures with an unknown number of components - an alternative to reversible jump methods. Ann. Statist. **28** (2000) 40–74.
54. Cappé, O., Robert, C.P., Rydén, T.: Reversible jump, birth-and-death and more general continuous time Markov chain Monte Carlo. J. R. Statist. Soc. B **65** (2003) 679–699.

Feature Selection for Dimensionality Reduction

Dunja Mladenić

Jožef Stefan Institute, Jamova 39, 1000 Ljubljana, Slovenia
Dunja.Mladenic@ijs.si
http://kt.ijs.si/Dunja

Abstract. Dimensionality reduction is a commonly used step in machine learning, especially when dealing with a high dimensional space of features. The original feature space is mapped onto a new, reduced dimensionally space. The dimensionality reduction is usually performed either by selecting a subset of the original dimensions or/and by constructing new dimensions. This paper deals with feature subset selection for dimensionality reduction in machine learning. We provide a brief overview of the feature subset selection techniques that are commonly used in machine learning. Detailed description is provided for feature subset selection as commonly used on text data. For illustration, we show performance of several methods on document categorization of real-world data.

1 Introduction

Machine learning can be used on different tasks that are often characterized by a high dimensional space of features [41]. Moreover, features used to describe learning examples are not necessarily all relevant and beneficial for the inductive learning task. Additionally, a high number of features may slow down the induction process while giving similar results as obtained with a much smaller feature subset. We can say that the main reasons for using dimensionality reduction in machine learning are: to improve the prediction performance, to improve learning efficiency, to provide faster predictors possibly requesting less information on the original data, to reduce complexity of the learned results and enable better understanding of the underlying process.

The original feature space is mapped onto a new, reduced dimensionality space and the original examples are then represented in the new space. The mapping is usually performed either by selecting a subset of the original features or/and by constructing new features [25]. Dimensionality reduction by selecting a subset of features does not involve any feature transformation, but rather concentrates on selecting features among the existing features. Dimensionality reduction by constructing new features can be performed by applying some methods from statistics that construct new features to be used instead of the original features (such as, principal components analysis or factor analysis [39]) or by using some background knowledge [20] for constructing new features usually to be used in addition to the original features. The background knowledge can be either general (such as, explicit functions calculating the value of new

C. Saunders et al. (Eds.): SLSFS 2005, LNCS 3940, pp. 84–102, 2006.

features based on the original features eg., sum of two features) or domain specific (such as, using user provided domain-specific functions for combining the original features into a new feature, using a language parser on text data to extract new features from text eg. noun phrases, using clustering of words on text data, see Section 3.1). Feature construction that adds new features can be followed by feature subset selection (the original feature set is first extended by the constructed features and then a subset of features is selected).

This paper focuses on dimensionality reduction using feature subset selection. We illustrate the effect of using different feature subset selection methods on the problem of text classification. We found feature selection in text classification [36, 45] especially interesting, as the number of features usually greatly exceeds the number of training documents (all the words occurring in the training documents are commonly used as features). Another common observation on text classification tasks is the high imbalance of the class distribution [46] with very few examples from the class of interest (eg., documents matching the user interest). We will see that there are feature selection methods that treat the class of interest differently from the other classes.

The rest of this paper is organised as follows: Section 2 gives a brief overview of the feature subset selection approaches commonly used in machine learning. Feature subset selection used on text data is described is details in Section 3. Section 4 illustrates the influence of feature subset selection to the system performance on document categorization. The presented feature selection methods are discussed in Section 5.

2 Feature Subset Selection in Machine Learning

The problem of selecting subset of the original features can be seen as an optimizations problem. The whole search space for optimization contains all possible subsets of features, meaning that its size is $\sum_{k=0}^{D} \binom{D}{k} = (1+1)^D = 2^D$, where D is the dimensionality (the number of features) and k is the size of the current feature subset. Different methods have been developed and used for feature subset selection in statistics, pattern recognition and machine learning, using different search strategies and evaluation functions. In machine learning, the following search strategies are commonly used for selecting a feature subset.

Forward selection — start with an empty set and greedily add features one at a time.

Backward elimination — start with a feature set containing all features and greedily remove features one at a time.

Forward stepwise selection — start with an empty set and greedily add or remove features one at a time.

Backward stepwise elimination — start with a feature set containing all features and greedily add or remove features one at a time.

Random mutation — start with a feature set containing randomly selected features, add or remove randomly selected feature one at a time and stop after a given number of iterations.

According to the literature, the approaches to feature subset selection can be divided into filters, wrappers and embedded approaches [25]. *Filters* perform feature subset selection based on some evaluation function that is independent of the learning algorithm that will be used later. *Wrappers* use the same machine learning algorithm that will be used later for modeling as a black-box for evaluating the feature subsets. *Embedded approaches* perform feature selection during the model generation (eg., decision trees), as opposite to the previous two approaches that perform it in a pre-processing phase. The *Simple filtering approach* is based on the filters, but it assumes the feature independence. It is used when dealing with very large number of features, such as in text data. In this way the solution quality is traded for the time needed to find the solution, justified by the large number of features.

2.1 Filtering Approach

In the filtering approach illustrated in Figure 1, a feature subset is selected independently of the learning method that will be applied on the examples represented using the selected feature subset. Here we briefly describe several feature subset selection algorithms used in machine learning that are based on the filtering approach.

Koller and Sahami [32] proposed an algorithm for feature subset selection that uses backward elimination to eliminate predefined number of features. The

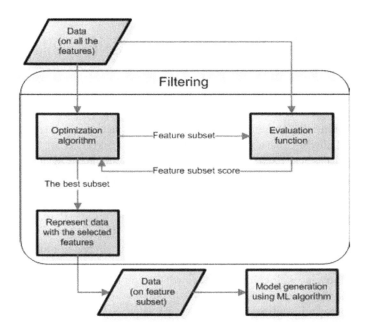

Fig. 1. Illustration of the filtering approach to feature subset selection. The evaluation function does not include characteristics of the machine learning algorithm that will use the selected features.

idea is to select a subset of features that keeps the class probability distribution as close as possible to the original distribution that is obtained using all the features. The algorithm starts with all the features and performs backward elimination to eliminate a predefined number of features. The evaluation function selects the next feature to be deleted based on the Cross entropy measure (see Section 3.1). For each feature the algorithm finds a subset of K features such that, the feature is approximated to be conditionally independent of the remaining features. Setting $K = 0$ results in a much faster algorithm that is equal to a simple filtering approach commonly used on text-data (see Section 3).

In the Relief algorithm [31] the main idea is to estimate quality of the features according to how well their values distinguish between examples that are similar. The feature score is calculated from a randomly selected subset of training examples, so that each example is used to calculate the difference in the distance from the nearest example of the same class and the nearest example of the different class. The nearest instances are found using the k-Nearest Neighbor algorithm. Some theoretical and empirical analysis of the algorithm and its extensions is provided in [55].

Almallim and Dietterich [3] developed several feature subset selection algorithms including a simple exhaustive search and algorithms that use different heuristics. They based their feature subset evaluation function on conflicts in class value occurring when two examples have the same values for all the selected features. In the first version of the feature subset selection algorithm called FOCUS, all the feature subsets of increasing size are evaluated until a *sufficient* subset is encountered. Feature subset Q is said to be sufficient if there are no *conflicts*. In other words, if there is no pair of examples that have different class values and the same values for all the features in Q. The successor of that algorithm FOCUS-2 prunes the search space, thus evaluating only promising subsets. Both algorithms assume the existence of a small set of features that form a solution and their usage on domains with a large number of features can be computationally infeasible. This is the reason why search heuristics are used in the next versions of the algorithm resulting in good but not necessarily optimal solutions. The idea is to use forward selection until a sufficient subset is encountered. The proposed heuristic algorithms differ in the evaluation function they use. One of the heuristic algorithms uses conditional class entropy of training examples when positioned into $2^{|Q|}$ groups such that the examples in each group have the same truth assignment to the features in Q. $Entropy(Q) = -\sum_{k=0}^{2^{|Q|}-1} P(Group_k) \sum_i P(C_i|Group_k) \log_2 P(C_i|Group_k)$, where Q is the current feature subset we are evaluating, $P(Group_k)$ is the probability of the k-th group of examples containing training examples that all have the same truth assignment to the features in Q, $P(C_i|Group_k)$ is the conditional probability of the i-th class value given the k-th group of examples. The search stops when the previously described class entropy equals 0. The other two heuristic algorithms they developed are based on the number of conflicts caused by representing examples using only selected features. The search stops when there are no conflicts (the sufficient feature subset is encountered).

Liu and Setiono [38] used random sampling to search the space of all possible feature subsets. The predetermined number of subsets is evaluated and the smallest having an inconsistency rate below the given threshold is selected. They define inconsistency rate as the average difference between the number of examples with the same values of all the selected features and the number of examples among them having the same, most frequent class value among these examples. Noise handling is enabled by setting the threshold to some positive value. If the threshold is set to 0, the evaluation is based on consistency check as used in the already described algorithm FOCUS [3].

Pfahringer [51] proposed using Minimum Description Length as an evaluation function of feature subset. In order to calculate Minimum Description Length of a feature subset, training examples are represented with a *simple decision table* that contains only features from the subset. He used a search that adds or removes features one at a time. The search starts with the randomly selected feature subset found to be the best among a fixed number of randomly selected feature subsets. He reported that the method performs at least as well as the wrapper approach applied on the simple decision tables and scales up better to a large number of training examples and features.

2.2   Wrapper Approach

The idea of the wrapper approach is to select a feature subset using the evaluation function based on the same learning algorithm that will be used later for learning [29] (see Figure 2 for illustration). Instead of using subset sufficiency, entropy or some other explicitly defined evaluation function, a kind of 'black box' function is used to guide the search. Namely, the evaluation function for each candidate feature subset returns the estimated quality of the model induced by the target learning algorithm using the feature subset. The idea is similar to automated model selection, where pruning parameters for decision tree induction are set by an optimization algorithm [42], [43]. This can result in a rather time consuming process, since for each candidate feature subset that is evaluated during the search, the target learning algorithm is usually applied several times (eg. 10 in case of using 10-fold cross validation for estimating the model quality). Here we briefly describe several feature subset selection algorithms developed in machine learning that are based on the wrapper approach.

Aha and Bankert [2] used a wrapper approach for feature subset selection in instance-based learning. They proposed a new search strategy that performs beam search using a kind of backward elimination. Namely, instead of starting with an empty feature subset, their search randomly selects a fixed number of feature subsets and starts with the best among them. Skalak [58] used a wrapper approach for feature subset selection and for selecting a subset of examples to be stored in instance-based learning. Instead of the deterministic search strategy used in [2] they used random mutation.

Caruana and Freitag [13] developed a wrapper for feature subset selection method for decision tree induction. They proposed a new search strategy that

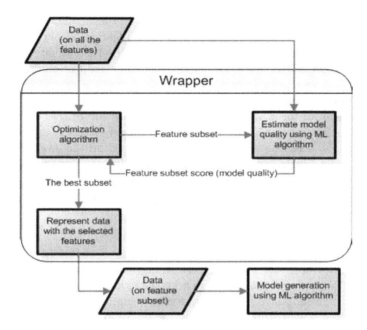

Fig. 2. Illustration of the wrapper approach to feature subset selection. The evaluation function is based on the same machine learning algorithm that will be used on the selected feature subset.

uses adding and removing features (backward stepwise elimination) and additionally at each step removes all the features that were not used in the decision tree induced for the evaluation of the current feature subset. Bala et al. [5] and Cherkauer and Shavlik [16] also used a wrapper approach for feature subset selection for decision tree induction but their method uses a genetic algorithm to perform the search. John et al. [29] defined the notion of strong and weak relevance of a feature and used a wrapper feature subset selection for decision tree induction.

Vafaie and De Jong [59] used a kind of wrapper approach where instead of using the same machine learning algorithm (AQ15 in this case) for evaluation and learning, the evaluation is performed using approximation of the model quality estimated by machine learning algorithm that selects feature subset which maximally separates classes using Euclidean distance. Their method uses a genetic algorithm to perform the search.

Feature subset selection can be performed using metric-based method for model selection. As in the above described wrapper approach, the feature subset is evaluated by applying some machine learning algorithm on the data represented only by the feature subset. However, in metric-based model selection the algorithm used for evaluation does not need to be the same as the algorithm to be used later on the feature subset. Also the data set used for comparing the two feature subsets can be unlabeled. The thesis is that models that overfit the data are likely to behave differently on training data than on the other data.

The feature subset is evaluated by inducing a model on the data represented by the feature subset and comparing the model classification output on the training data and on the unlabeled data. The feature subset is evaluated relative to the currently best feature subset by comparing the classification output of the two models (eg., average difference in the prediction of the first model from the prediction of the second model).

An extension of the wrapper approach to feature subset selection has been proposed [9] that combines the metric-based model selection and the cross-validation model selection. The combination is based on the level of their disagreement on testing examples, so that the higher disagreement means the lower trust to the cross-validation model. The intuition behind is that cross-validation mainly provides good results but has high variance and should benefit from a combination with some other model selection approach with lower variance.

2.3 Embedded Approach

In the embedded approach to feature selection, the feature selection and learning are interleaved (see Figure 3). A well known example is decision tree induction [52], where a simplified filtering approach assuming feature independence is used to find the best feature for splitting the data into subsets. The procedure is repeated recursively on the subsets until some stopping criterion is satisfied. The output is a model that uses only a subset of features (those that appear in the nodes of decision tree). Notice that the embedded approach assumes that our learning algorithm already has feature selection interleaved with learning.

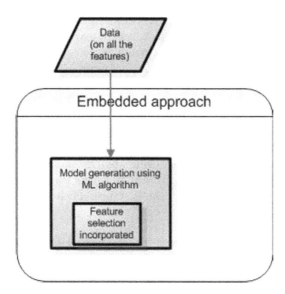

Fig. 3. Illustration of the embedded approach to feature subset selection. Feature selection is an integral part of the machine learning algorithm that is use for model generation.

Feature selection that is an integral part of the model generation as proposed in [50] uses a fast gradient-based heuristic to find the most promising feature for improving the existing model at each iteration step of the incremental optimization of the model.

Embedded feature selection based on training a non-linear SVM is proposed in [53]. The features are selected using backward elimination based on the criteria derived from generalization error bounds of the SVM theory: the weight vector norm or alternatively using upper bounds of the leave-one-out error. The idea behind the approach is that features that are relevant to the concept should affect the generalization error bound more than irrelevant features.

The embedded approach to feature subset selection can be used as a part of a filtering approach. This can be useful if we want to use a learning algorithm that does not have feature selection embedded (as proposed for the Nearest Neighbor algorithm in [12]) or we run the algorithm with embedded feature selection on a data sample and re-run it on the complete dataset but with the reduced feature set (as proposed for text data in [11]). Namely, instead of using the model that has embedded feature selection, only the features that appear in the model are taken to form the selected feature subset. An example is the algorithm proposed in [12] that uses decision tree induction for feature subset selection. The idea is that only the features that appear in the induced decision tree are selected for learning using Nearest Neighbor algorithm. In this algorithm the optimization process is based on a feature selection measure (in this case information gain ratio) that is used in the greedy search to induce decision tree.

2.4   Discussion

Feature subset selection can be enhanced to provide additional information to system by introducing a mapping from the selected features to the discarded features in the multitask learning setting [14]. Experiments on synthetic regression and classification problems and real-world medical data have shown improvements in the system performance. Namely, the features that harm performance if used as inputs were found to improve performance if used as an additional output. The idea is that transfer of information is occurring inside the model, when in addition to the original output it models also that additional output consisting of the discarded features.

Multitask learning was used for SVM [27], where the kernel and learning parameters were shared between different tasks (SVM models). The method is applicable when several classification tasks with differently labeled data set share common input space.

3   Feature Selection on Text Data

Most of the feature subset selection approaches used in machine learning are not designed for the situations with a large number of features. The usual way of learning on text defines a feature for each word that occurs in the training documents. This can easily result in several tens of thousands of features. Most

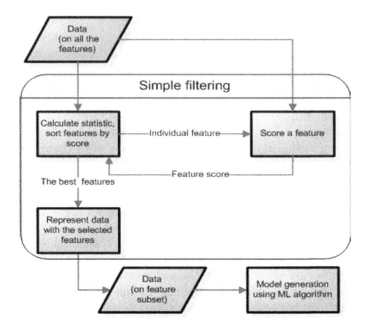

Fig. 4. Illustration of the approach to feature subset selection commonly used on text data. Each feature is scored independently of other features. Features are sorted according to their score and the first k features are selected to form the solution feature subset.

methods for feature subset selection that are used on text are very simple compared to the described methods developed in machine learning. Basically, some evaluation function that is applied to a single feature is used as described in Section 3.1. All the features are independently evaluated, a score is assigned to each of them and the features are sorted according to the assigned score. Then, a predefined number of the best features is taken to form the final feature subset (see Figure 4).

3.1   Feature Scoring in Text Domains

Scoring of individual features is usually performed in a supervised way, for instance, measuring mutual information between the feature and the class. However, there are some feature scoring measure that ignore the class information, such as scoring by the number of documents that contain a particular word.

Different scoring measures have been proposed on text data. *Information gain* used in decision tree induction [52] was reported to work well as feature scoring measure on text data [60] in some domains (Reuters-22173, a subset of MEDLINE) where a multiclass problem was addressed using k-Neareast Neighbor algorithm; while in others using Naive Bayesian classifier on a binary classification problem (sub domains of Yahoo!) it almost completely failed [47]. This difference in performance can be partially attributed to the classification algorithm and domain characteristics. In the domains having unbalanced class distribution, most of the features highly scored by Information gain will be representative of the majority class value, as the measure is symmetric in class values and in feature values (see the formula below). In binary classification problems with highly unbalanced class distribution, the Naive Bayesian classifier needs some evidence (features) of the minority class value unless it will by default classifying all the documents to the majority class value.

It is interesting to notice that *Information gain* takes into account all values for each feature (in the case of text classification two values: occurs or does not occur in a document), as opposite to the most other, more successful, feature scoring measures used on text that count only the feature occurrences. *Expected cross entropy* as used on text data ([33], [47]) is similar to *Information gain*, but instead of calculating the average over all the possible feature values, only the value denoting that the feature occurred in a document is used. Experiments on document categorization into a hierarchical topic taxonomy [47] have show that this significantly improves the performance. *Expected cross entropy* is related to *Information gain* as follows: $InfGain(F) = CrossEntropyTxt(F) + CrossEntropyTxt(\overline{F})$, where F is a binary feature (usually representing a word's occurrence). *Cross entropy for text*, *Frequency*, *Mutual information* are measures that were reported to work well on text data [60].

Odds ratio was reported to outperform many other measures [47] in combination with Naive Bayes used for document categorization on data with highly imbalanced class distribution. A characteristic of Naive Bayes used for text classification is that, once the model has been generated, the classification is based on the features that occur in a document to be classified. This means that an empty document will be classified into the majority class. Consequently, having a highly imbalanced class distribution, if we want to identify documents from the under-represented class value we need to have a model sensitive to the features that occur in such documents. If most of the selected features are representative for the majority class value, the documents from other classes will be almost empty (represented using only the selected features).

Experimental comparison of different feature selection measures in combination with the Support Vector Machine classification algorithm (SVM) on a Reuters-2000 data set [11] has shown that using all or almost all the features yields the best performance. The same finding was confirmed in experimental evaluation of different feature selection measures on a number of text classification problems [22]. In addition, in [22] a new feature selection measure was introduced $Bi - NormalSeparation$ that was reported to improve the performance of SVM especially on the problems where the class distribution is highly imbalanced. Interestingly, they also report that Information gain is outperforming the other tested measures in the situation when using only a small number of selected features (20-50 features).

Another feature selection method for text data called *Fisher index* was proposed as a part of document retrieval system based on organizing large text databases into hierarchical topic taxonomies [15]. Similar to [45], for each internal node in the topic taxonomy a separate feature subset is calculated and used to build a Naive Bayesian classifier for that node. In that way, the feature set used in the classification of each node is relatively small and is changed with context. In contrast to the work reported in [45, 47], and similar to the work of [33], a multilevel classifier is designed and used to classify a document to a leaf node. Namely, a new document is classified from the top of the taxonomy and based on the classification outcome in the inner nodes, proceeds down the taxonomy.

What follows are formulas of the described scoring measures.

$$InfGain(F) = P(F) \sum_i P(C_i|F) \log \frac{P(C_i|F)}{P(C_i)} + P(\overline{F}) \sum_i P(C_i|\overline{F}) \log \frac{P(C_i|\overline{F})}{P(C_i)}$$

$$CrossEntropyTxt(F) = P(F) \sum_i P(C_i|F) \log \frac{P(C_i|F)}{P(C_i)}$$

$$MutualInfoTxt(F) = \sum_i P(C_i) \log \frac{P(F|C_i)}{P(F)}$$

$$Freq(F) = TF(F)$$

$$OddsRatio(F) = \log \frac{P(F|pos)(1 - P(F|neg))}{(1 - P(F|pos))P(F|neg)}$$

$$Bi - NormalSeparation(F) = Z^{-1}(P(F|pos)) - Z^{-1}(P(F|neg))$$

$$FisherIndexTxt(F) = \frac{\sum_{pos,neg}(P(F|pos) - P(F|neg))^2}{\sum_{C_i \in pos,neg} \frac{1}{|C_i|} \sum_{d \in C_i}(n(F,d) - P(F|C_i))^2}$$

Where $P(F)$ is the probability that feature F occurred, $\overline{F}$ means that the feature does not occur, $P(C_i)$ in the probability of the i-th class value, $P(C_i|F)$ is the conditional probability of the i-th class value given that feature F occurred, $P(F|C_i)$ is the conditional probability of feature occurrence given the i-th class value, $P(F|pos)$ is the conditional probability of feature F occurring given the class value 'positive', $P(F|neg)$ is the conditional probability of feature F occurring given the class value 'negative', $TF(F)$ is term frequency (the number of times feature F occurred), $Z^{-1}(x)$ is the standard Normal distribution's inverse cumulative probability function (z-score), $|C_i|$ is the number of documents in class C_i and $n(F,d)$ is 1 if the document d contains feature F and 0 otherwise.

As already highlighted in text classification most of the feature selection methods evaluate each feature independently. A more sophisticated approach is proposed in [11] where a linear SVM is first trained using all the features and then, the induced model is used to score the features (weight assigned to each feature in the normal to the induced hyper plane is used as a feature score). Experimental evaluation using that feature selection in combination with SVM, Perceptron and Naive Bayes has shown that the best performance is achieved by SVM when using almost all the features. The experiments have confirmed the previous findings [46] on feature subset selection improving the performance of Naive Bayes, but the overall performance is lower than using SVM on all the features. Similar as in [11], feature selection was performed using a linear SVM to rank the features in [8]. However, the experiments in [8] were performed on a regression problem and the final model was induced using a nonlinear SVM. The feature selection was shown to improve the performance.

Distributional clustering of words with an agglomerative approach (words are viewed as distributions over document categories) is used for dimensionality reduction via feature construction [10] that preserves the mutual information between the features as much as possible. This representation was shown to achieve comparable or better results than the bag-of-words document representation using feature selection based on Mutual Information for Text; a linear SVM was used as the classifier. Related approach also based on preserving the mutual information between the features [24] finds new dimensions by using an iterative projection algorithm instead of clustering. It was shown to achieve performance comparable to the bag-of-words representation with all the original features, using significantly less features (eg., on one dataset four constructed features achieved 98% of performance of 500 original features) using the linear SVM classifier.

Divisive clustering for feature construction [18] was shown to outperform distributional clustering when used for dimensionality reduction on text data. The approach uses Kullback-Leibler divergence as a distance function and minimizes within-cluster divergence while maximizing between-cluster divergence. Experiments on two datasets have shown that this dimensionality reduction slightly improves the performance of Naive Bayes (compared to using all the original features), outperforming the agglomerative clustering of words combined with Naive Bayes and achieving considerably higher classification accuracy for the same number of features than feature subset selection using Information Gain or Mutual Information (in combination with Naive Bayes or SVM).

4   Illustration of Feature Selection Influence to Performance

We now illustrate the feature selection influence on the performance of document categorization into a hierarchy of Web documents. The experimental results report the influence of using different sizes of the feature subset and different scoring measures. The results of the experiments are summarized based on the work reported in [47]. Briefly, the experiments are performed so that the documents are represented as feature-vectors with features being sequences of n consecutive words (n-grams). A separate classifier for each of the hierarchy nodes is induced using a Naive Bayesian classifier on text data in a similar way as described in [28], [41] or [44]. A set of positive and negative examples is constructed for each hierarchy node based on the examples in the node and weighted examples from its sub-nodes. The formula used to predict probability of class value C for a given document Doc is the following: $P(C|Doc) = \frac{P(C)\Pi_j P(F_j|C)^{TF(F_j,Doc)}}{\sum_i P(C_i)\Pi_l P(F_l|C_i)^{TF(F_l,Doc)}}$, where the product goes over all the features used in the representation of documents (in our case all selected features), $P(C)$ is the probability of class value C, $P(F_j|C)$ is the conditional probability that feature F_j occurs in a document given the class value C calculated using Laplace probability estimate. $TF(F_j, Doc)$ is the frequency of feature F_j in document Doc. Since $TF(F_j, Doc) = 0$ for all the words that are not contained in document Doc, the predicted class probability can be calculated by taking into account only words (in our case selected features) contained in document Doc as follows:

$$P(C|Doc) = \frac{P(C)\Pi_{F \in Doc}P(F|C)^{TF(F,Doc)}}{\sum_i P(C_i)\Pi_{F_l \in Doc}P(F_l|C_i)^{TF(F_l,Doc)}} \tag{1}$$

4.1   Experimental Results

Results of 5-fold cross validation are evaluated using standard document classification measures, precision and recall and reported in Table 1. Precision is

Table 1. Comparison of different feature selection measures for document categorization on domains formed from Yahoo! hierarchy based on [47]. We observe performance (F2-measure) for the best performing number of features. Additionally, we give values of Precision and Recall. Feature scoring measures are sorted according to their performance in F2-measure. Report are the average values and standard errors.

Domain name	Scoring measure	Average on category prediction		
		F2-measure	Precision	Recall
Ent.	**Odds ratio**	**0.30** ± 0.003	**0.41** ± 0.004	**0.34** ± 0.003
	Term frequency	0.27 ± 0.003	0.38 ± 0.003	0.34 ± 0.003
	Mutual info. for text	0.23 ± 0.004	0.57 ± 0.006	0.29 ± 0.007
	Information gain	0.20 ± 0.003	0.87 ± 0.002	0.17 ± 0.002
	Cross entropy for text	0.18 ± 0.002	0.29 ± 0.005	0.28 ± 0.005
	Random	0.001± 0.0002	0.99 ± 0.007	0.001±0.0002
Arts.	**Odds ratio**	**0.32** ± 0.004	**0.36** ± 0.005	**0.38** ± 0.004
	Term frequency	0.29 ± 0.003	0.43 ± 0.004	0.34 ± 0.003
	Mutual info. for text	0.25 ± 0.005	0.56 ± 0.006	0.31 ± 0.007
	Cross entropy for text	0.22 ± 0.003	0.27 ± 0.005	0.32 ± 0.006
	Information gain	0.17 ± 0.002	0.93 ± 0.003	0.15 ± 0.002
	Random	0.001± 0.0003	0.99 ± 0.006	0.001±0.0002
Comp.	**Odds ratio**	**0.33** ± 0.002	**0.36** ± 0.009	**0.57** ± 0.005
	Term frequency	0.26 ± 0.002	0.45 ± 0.003	0.27 ± 0.003
	Mutual info. for text	0.24 ± 0.004	0.60 ± 0.006	0.26 ± 0.006
	Cross entropy for text	0.21 ± 0.004	0.28 ± 0.004	0.27 ± 0.002
	Information gain	0.14 ± 0.006	0.94 ± 0.004	0.12 ± 0.005
	Random	0.001± 0.0002	0.99 ± 0.001	0.001± 0.0002
Ref.	**Odds ratio**	**0.42** ± 0.009	**0.46** ± 0.009	**0.51** ± 0.012
	Mutual info. for text	0.32 ± 0.005	0.69 ± 0.015	0.32 ± 0.006
	Term frequency	0.26 ± 0.070	0.72 ± 0.007	0.26 ± 0.007
	Cross entropy for text	0.22 ± 0.005	0.50 ± 0.003	0.23 ± 0.005
	Information gain	0.16 ± 0.002	0.99 ± 0.002	0.14 ± 0.002
	Random	0.04 ± 0.005	0.99 ± 0.001	0.04 ± 0.004

defined as the proportion of correctly classified examples in the set of all the examples that are assigned the target class value (similar to classification accuracy measured for one class value), while recall is defined as the proportion of correctly classified examples out of all the examples having the target class value. Precision and recall are complementary and should be maximized. We also report a value of $F - measure$ that is a combination of Precision and Recall commonly used in information retrieval, where the relative importance of each is expressed with the value of parameter β. $F_\beta = \frac{(1+\beta^2)Precision \times Recall}{\beta^2 Precision + Recall}$.

We also report on the influence of the number of selected features, as the reported experiments were performed for different values of the predefined number of features to be selected. Influence of the feature subset size can be seen from the performance graphs in Figure 5. It is evident that in the reported experiments reducing the feature subset improves the overall performance of the system.

It can be seen that Odds ratio is among the best performing measures, while Information gain is one of the worst performing measures together with Random scoring. Except Random scoring, all other measures achieve the best results for vector size around 1, meaning that the feature subset is approximately of the size of positive documents' vocabulary.

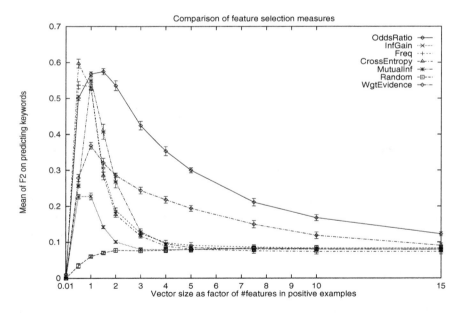

Fig. 5. Comparison of F-measure values for different feature scoring measures on domain *'Reference'* defined from the Yahoo hierarchy. The problem is assignment of document into a node of the existing document hierarchy. Given results are for pruning setting = (0.7, 3) and probability threshold 0.95. We give mean and standard error of F2-measure.

5 Discussion

The paper provides an overview of dimensionality reduction by feature subset selection techniques that are commonly used in machine learning. We can say that the main reasons for using dimensionality reduction in machine learning are: to improve the prediction performance, to improve learning efficiency, to provide faster predictors possibly requesting less information on the original data, to reduce complexity of the learned results and enable better understanding of the underlying process.

Dimensionality reduction in machine learning is performed usually either by feature selection or feature construction. Selecting a subset of features does not involve any feature transformation, but rather concentrates on selecting features among the existing features. Constructing new features can be performed by applying some methods from statistics that construct new features to be used instead of the original features or by using some background knowledge for constructing new features usually to be used in addition to the original features. Actually, feature construction that adds new features can be followed by feature subset selection. The approaches to feature subset selection can be divided into filters, wrappers and embedded approaches [25]. *Filters* perform feature subset selection based on some evaluation function that is independent of the learning algorithm that will be used later. *Wrappers* use the same machine learning algorithm that will be used later for modeling as a black-box for evaluating the feature subsets. *Embedded approaches* perform feature selection during the model generation, as opposite to the previous two approaches that perform feature selection as pre-processing.

We illustrate the effect of using different feature subset selection methods on the problem of text classification. A variant of the filtering approach assuming feature independence, *Simple filtering approach*, is commonly used when dealing with very large number of features, such as in text data. We found feature selection in text classification especially interesting, as the number of features usually greatly exceeds the number of training documents. Another common observation on text classification tasks is the high imbalance of the class distribution with very few positive examples. For illustration, real-world data is used to show performance of several feature selection methods on the problem of document categorization. It can be seen that the performance of some classification algorithms, such as Naive Bayes, greatly depends on the selected feature subset. On the other hand, there are algorithms, such as SVM, that have rather stable performance if the feature subset is sufficiently large (eg., containing 50% of the original features). In that light, one should be aware of the algorithm's sensitivity to the feature set prior to applying any feature subset selection. It was also found that some datasets are more sensitive to feature subset selection than others. Our hope is that this paper contributes by giving an overview of the approaches and illustrating potential influence of the feature subset selection on the classification performance.

References

1. Agrawal, R., Mannila, H., Srikant, R., Toivonen, H., & Verkamo, A.I. (1996). Fast Discovery of Association Rules. In U.M. Fayyad, G. Piatetsky-Shapiro, P. Smyth, R. Uthurusamy (Eds.), *Advances in Knowledge Discovery and Data Mining* AAAI Press/The MIT Press.

2. Aha, D.W., & Bankert, R.L. (1994). Feature selection for case-based classification of cloud types: An empirical comparison, *Proceedings of the AAAI'94 Workshop on Case-Based Reasoning* (pp. 106-112). Seattle, WA:AAAI Press.

3. Almuallin, H., & Dietterich, T.G. (1991). Efficient algorithms for identifying relevant features. *Proceedings of the Ninth Canadian Conference on Artificial Intelligence* (pp. 38-45). Vancouver, BC:Morgan Kaufmann.

4. Apté, C., Damerau, F., & Weiss, S.M. (1994). Toward Language Independent Automated Learning of Text Categorization Models. *Proceedings of the 7th Annual International ACM-SIGIR Conference on Research and Development in Information Retrieval.* Dubline.

5. Bala, J., Huang, J., & Vafaie, H. (1995). Hybrid Learning Using Genetis Algorithms and Decision Trees for Pattern Classification. *Proceedings of the 14th International Joint Conference on Artificial Intelligence IJCAI-95* (pp. 719-724), Montreal, Quebec.

6. Balabanović, M., & Shoham, Y. (1995). Learning Information Retrieval Agents: Experiments with Automated Web Browsing. *Proceedings of the AAAI 1995 Spring Symposium on Information Gathering from Heterogeneous, Distributed Environments.* Stanford.

7. Berry, M.W., Dumais, S.T., & OBrein, G.W. (1995). Using linear algebra for intelligent information retrieval. *SIAM Review, 4,* (Vol. 37), 573-595.

8. Bi, J., Bennett, K.P., Embrechts, M., Breneman, C.M., Song, M. (2003). Dimensionality Reduction via Sparse Support Vector Machines. *Journal of Machine Learning Research* 3, 1229-1243.

9. Bengio, Y., Chapados, N. (2003). Extensions to Metric-Based Model Selection. *Journal of Machine Learning Research,* 3, 1209-1227.

10. Bekkerman, R., El-Yaniv, R., Tishby, N., Winter, Y. (2003). Distributional Word Clusters vs. Words for Text Categorization. Journal of Machine Learning Research, 3, 1183-1208.

11. Brank, J., Grobelnik, M., Milić-Frayling, N., Mladenić, D. (2002). Feature selection using support vector machines. In *Data mining III* (Zanasi, A., ed.). Southampton: WIT, 261-273.

12. Cardie, C. (1993). Using Decision Trees to Improve Case-Based Learning. *Proceedings of the 10th International Conference on Machine Learning ICML93* (pp. 25-32).

13. Caruana, R., & Freitag, D. (1994). Greedy Attribute Selection. *Proceedings of the 11th International Conference on Machine Learning ICML94* (pp. 28-26).

14. Caruana, R., de Sa, V.R. (2003). Benefitting from the Variables that Variable Selection Discards. *Journal of Machine Learning Research,* 3, 1245-1264.

15. Chakrabarti, S., Dom, B., Agrawal, R., Raghavan, P. (1998). Scalable feature selection, classification and signature generation for organizing large text databases into hierarchical topic taxonomies. *The VLDB Journal,* 7, 163-178, Spinger-Verlag.

16. Cherkauer, K.J., & Shavlik, J.W. (1996). Growing simpler decision trees to facilitate knowledge discovery, *Proceedings of the Second International Conference on Knowledge Discovery and Data Mining KDD-96* (pp. 315-318). Portland, OR:AAAI Press.

17. Cohen, W.W. (1995). Learning to Classify English Text with ILP Methods. *Proceedings of the Workshop on Inductive Logic Programming.* Leuven.
18. Dhillon, I., Mallela, S., Kumar, R. (2003). A Divisive Information-Theoretic Feature Clustering Algorithm for Text Classification. *Journal of Machine Learning Research,* 3, 1265-1287.
19. Domingos, P., & Pazzani, M. (1997). On the Optimality of the Simple Bayesian Classifier under Zero-One Loss. *Machine Learning, 29,* 103-130. Kluwer Academic Publishers.
20. Dzeroski, S., Lavrac, N. (2001). An introduction to inductive logic programming. In: Relational data mining, Dzeroski, Lavrac (eds.) Berlin:Springer, 2001, pp. 48-73.
21. Filo, D., & Yang, J. (1997). Yahoo! Inc. http://www.yahoo.com/docs/pr/
22. Forman, G. (2003). An Extensive Empirical Study of Feature Selection Metrics for Text Classification, *Journal of Machine Learning Research* 3, 1289-1305.
23. Frakes W.B., & Baeza-Yates R., (Eds.) (1992). Information Retrieval: Data Structures & Algorithms. Englewood Cliffs: Prentice Hall.
24. Globerson, A., Tishby, N. (2003). Sufficient Dimensionality Reduction, Journal of Machine Learning Research, 3, 1307-1331.
25. Guyon, I., Elisseeff, A. (2003). An Introduction to Variable and Feature Selection. *Journal of Machine Learning Research* 3, pp. 1157-1182.
26. Grobelnik, M., & Mladenić, D. (1998). Learning Machine: design and implementation. *Technical Report IJS-DP-7824.* Department for Intelligent Systems, J.Stefan Institute, Slovenia.
27. Jebara, T. (2004). Multi-Task Feature and Kernel Selection for SVMs. Proceedings of the International Conference on Machine Learning, ICML-2004.
28. Joachims, T. (1997). A Probabilistic Analysis of the Rocchio Algorithm with TFIDF for Text Categorization. *Proceedings of the 14th International Conference on Machine Learning ICML97* (pp. 143-151).
29. John, G.H., Kohavi, R., & Pfleger, K. (1994). Irrelevant Features and the Subset Selection Problem. *Proceedings of the 11th International Conference on Machine Learning ICML94* (pp. 121-129).
30. Kindo, T., Yoshida, H., Morimoto, T., & Watanabe, T. (1997). Adaptive Personal Information Filtering System that Organizes Personal Profiles Automatically. *Proceedings of the 15th International Joint Conference on Artificial Intelligence IJCAI-97* (pp. 716-721).
31. Kira, K., Rendell, L.A. (1992). The feature selection problem: Traditional methods and a new algorithm. *Proceedings of the Ninth National Conference on Artificial Intelligence AAAI-92* (pp.129-134). AAAI Press/The MIT Press.
32. Koller, D., & Sahami, M. (1996). Toward optimal feature selection. *Proceedings of the 13th International Conference on Machine Learning ICML96* (pp. 284-292).
33. Koller, D., & Sahami, M. (1997). Hierarchically classifying documents using very few words. *Proceedings of the 14th International Conference on Machine Learning ICML97* (pp. 170-178).
34. Kononenko, I. (1995). On biases estimating multi-valued attributes. *Proceedings of the 14th International Joint Conference on Artificial Intelligence IJCAI-95* (pp. 1034-1040).
35. Lam, W., Low, K.F., & Ho, C.Y. (1997). Using Bayesian Network Induction Approach for Text Categorization. *Proceedings of the 15th International Joint Conference on Artificial Intelligence IJCAI97* (pp. 745-750).

36. Lewis, D.D. (1992). Feature Selection and Feature Extraction for Text Catego-
    rization. *Proceedings of Speech and Natural Language workshop* (pp. 212-217).
    Harriman, CA:Morgan Kaufmann.
37. Lewis, D.D. (1995). Evaluating and optimizing autonomous text classification
    systems. *Proceedings of the 18th Annual International ACM-SIGIR Conference on
    Recsearch and Development in Information Retrieval* (pp.246-254).
38. Liu, H., & Setiono, R. (1996). A probabilistic approach to feature selection - A filter
    solution. *Proceedings of the 13th International Conference on Machine Learning
    ICML'97* (pp. 319-327). Bari.
39. Manly, B.F.J. (1994). Multivariate Statistical Methods - a primer (Second ed.).
    Chapman & Hall.
40. Mansuripur, M. (1987). *Introduction to Information Theory.* Prentice-Hall.
41. Mitchell, T.M. (1997). Machine Learning. The McGraw-Hill Companies, Inc..
42. Mladenić, D. (1995). Automated model selection. *Proceedings of the MLNet famil-
    iarisation workshop: Knowledge level modelling and machine learning, ECML 95,*
    Heraklion.
43. Mladenić, D. (1995). Domain-Tailored Machine Learning. *M.Sc. Thesis*, Faculty of
    computer and information science, University of Ljubljan, Slovenia.
44. Mladenić, D. (1996). Personal WebWatcher: Implementation and Design. *Techni-
    cal Report IJS-DP-7472.* Department for Intelligent Systems, J.Stefan Institute,
    Slovenia.
45. Mladenić, D. (1998). Feature subset selection in text-learning. *Proceedings of the
    10th European Conference on Machine Learning ECML98.*
46. Mladenić, D., Grobelnik, M. (1999). Feature selection for unbalanced class distri-
    bution and Naive Bayes. In *Proceedings of the Sixteenth International Conference
    on Machine Learning (ICML-1999)*, 258-267, San Francisco: M. Kaufmann.
47. Mladenić, D., Grobelnik, M. (2003). Feature selection on hierarchy of web docu-
    ments. *Journal of Decision support systems*, 35, 45-87.
48. Mladenić, D., & Grobelnik, M. (2004). Mapping documents onto web page ontology.
    In: Web mining : from web to semantic web (Berendt, B., Hotho, A., Mladenic, D.,
    Someren, M.W. Van, Spiliopoulou, M., Stumme, G., eds.), Lecture notes in artificial
    inteligence, Lecture notes in computer science, vol. 3209, Berlin; Heidelberg; New
    York: Springer, 2004, 77-96.
49. Pazzani, M., & Billsus, D. (1997). Learning and Revising User Profiles: The Iden-
    tification of Interesting Web Sites. *Machine Learning, 27,* 313-331.
50. Perkins, S., Lacker, K., Theiler, J. (2003). Grafting: Fast, Incremental Feature
    Selection by Gradient Descent in Function Space. *Journal of Machine Learning
    Research*, 3, 1333-1356.
51. Pfahringer, B. (1995). Compression-Based Feature Subset Selection. In P. Turney
    (Ed.), *Proceedings of the IJCAI-95 Workshop on Data Engineering for Inductive
    Learning*, Workshop Program Working Notes, Montreal, Canada, 1995.
52. Quinlan, J.R. (1993). Constructing Decision Tree. In *C4.5: Programs for Machine
    Learning.* Morgan Kaufman Publishers.
53. Rakotomamonjy, A. (2003). Variable Selection Using SVM-based Criteria *Journal
    of Machine Learning Research*, 3, 1357-1370
54. van Rijsbergen, C.J,. Harper, D.J., & Porter, M.F. (1981). The selection of good
    search terms. *Information Processing & Management, 17,* 77-91.
55. Robnik-Sikonja, M., Kononenko, I. (2003). Theoretical and Empirical Analysis of
    ReliefF and RReliefF. *Machine Learning*, 53, 23-69, Kluwer Academic Publishers.
56. Shapiro, A. (1987). *Structured induction in expert systems.* Addison-Wesley.

57. Shaw Jr, W.M. (1995). Term-relevance computations and perfect retrieval performance. *Information Processing & Management, 31(4)* 491-498.
58. Skalak, D.B. (1994). Prototype and Feature Selection by Sampling and Random Mutation Hill Climbing Algorithms. *Proceedings of the 11th International Conference on Machine Learning ICML94* (pp. 293-301).
59. Vafaie, H., & De Jong, K. (1993). Robust feature selection algorithms. *Proceedings of the 5th IEEE International Conference on Tools for Artificial Intelligence* (pp. 356-363). Boston, MA: IEEE Press.
60. Yang, Y., & Pedersen, J.O. (1997). A Comparative Study on Feature Selection in Text Categorization. *Proceedings of the 14th International Conference on Machine Learning ICML-97* (pp. 412-420).

Auxiliary Variational Information Maximization for Dimensionality Reduction

Felix Agakov[1] and David Barber[2]

[1] University of Edinburgh, 5 Forrest Hill, EH1 2QL Edinburgh, UK
felixa@inf.ed.ac.uk
www.anc.ed.ac.uk
[2] IDIAP, Rue du Simplon 4, CH-1920 Martigny, Switzerland
www.idiap.ch

Abstract. Mutual Information (MI) is a long studied measure of information content, and many attempts to apply it to feature extraction and stochastic coding have been made. However, in general MI is computationally intractable to evaluate, and most previous studies redefine the criterion in forms of approximations. Recently we described properties of a simple lower bound on MI, and discussed its links to some of the popular dimensionality reduction techniques [1]. Here we introduce a richer family of *auxiliary variational* bounds on MI, which generalizes our previous approximations. Our specific focus then is on applying the bound to extracting informative lower-dimensional projections in the presence of irreducible Gaussian noise. We show that our method produces significantly tighter bounds than the well-known *as-if Gaussian* approximations of MI. We also show that the auxiliary variable method may help to significantly improve on reconstructions from noisy lower-dimensional projections.

1 Introduction

One of the principal goals of dimensionality reduction is to produce a lower-dimensional representation y of a high-dimensional source vector x, so that the useful information contained in the source data is not lost. If it is not known a priori which coordinates of x may be relevant for a specific task, it is sensible to maximize the amount of information which y contains about all the coordinates, for all possible x's. The fundamental measure in this context is the mutual information

$$I(\mathsf{x}, \mathsf{y}) \equiv H(\mathsf{x}) - H(\mathsf{x}|\mathsf{y}), \qquad (1)$$

which indicates the decrease of uncertainty in x due to the knowledge of y. Here $H(\mathsf{x}) \equiv -\langle \log p(\mathsf{x}) \rangle_{p(\mathsf{x})}$ and $H(\mathsf{x}|\mathsf{y}) \equiv -\langle \log p(\mathsf{x}|\mathsf{y}) \rangle_{p(\mathsf{x},\mathsf{y})}$ are marginal and conditional entropies respectively, and the angled brackets represent averages over all variables contained within the brackets.

The principled information-theoretic approach to dimensionality reduction maximizes (1) with respect to parameters of the *encoder* $p(\mathsf{y}|\mathsf{x})$, under the assumption that $p(\mathsf{x})$ is fixed (typically given by the empirical distribution). It is

C. Saunders et al. (Eds.): SLSFS 2005, LNCS 3940, pp. 103–114, 2006.

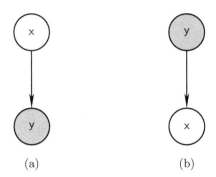

Fig. 1. Generative and encoder models. (a): An encoder model $\mathcal{M}_I \overset{\text{def}}{=} \tilde{p}(\mathsf{x})p(\mathsf{y}|\mathsf{x})$ trained by maximizing the mutual information $I(\mathsf{x},\mathsf{y})$ (b): a generative model $\mathcal{M}_L \overset{\text{def}}{=} p(\mathsf{y})p(\mathsf{x}|\mathsf{y})$ trained by maximizing the log-likelihood $\langle \log p(\mathsf{x}) \rangle_{\tilde{p}(\mathsf{x})}$. Here $\tilde{p}(\mathsf{x})$ is the empirical distribution, the shaded nodes indicate the hidden representations y, and we have assumed that the data patterns x are i.i.d.

easy to see that if the reduced dimension $|\mathsf{y}|$ ($|\mathsf{y}| < |\mathsf{x}|$) is still large, the exact evaluation of $I(\mathsf{x},\mathsf{y})$ is in general computationally intractable. The key difficulty lies in the computation of the conditional entropy $H(\mathsf{x}|\mathsf{y})$, which is tractable only in a few special cases. Typically, the standard techniques assume that $p(\mathsf{x},\mathsf{y})$ is jointly Gaussian (so that $I(\mathsf{x},\mathsf{y})$ has a closed analytical form), or the channels are deterministic and invertible [2], [3] (which may be related to the noiseless square ICA case). Alternatively, it is sometimes assumed that the output spaces are very low-dimensional, so that integration over y in the computation of $I(\mathsf{x},\mathsf{y})$ may be performed numerically). Clearly, these assumptions may be too restrictive for many subspace selection applications. Other existing methods suggest to optimize alternative objective functions (e.g. approximations of $I(\mathsf{x},\mathsf{y})$ based on the *Fisher Information* criterion [4]), which, however, do not retain proper bounds on $I(\mathsf{x},\mathsf{y})$ and may often lead to numerical instabilities when applied to learning undercomplete representations [5].

1.1   Encoder *vs* Generative Models

A principal motivation for applying information theoretic techniques for stochastic subspace selection and dimensionality reduction is the general intuition that the unknown compressed representations should be predictive about the higher-dimensional data. Additionally, we note that the information-maximizing framework of encoder models is particularly convenient for addressing problems of *constrained* dimensionality reduction [6], as by parameterizing $p(\mathsf{y}|\mathsf{x})$ we may easily impose *explicit* constraints on the projection to a lower-dimensional space (see Fig. 1 (a)). This is in contrast to generative latent variable models (Fig. 1 (b)) commonly used for probabilistic dimensionality reduction (e.g. [7], [8]), where the probabilistic projection to the latent space $p(\mathsf{y}|\mathsf{x}) \propto p(\mathsf{y})p(\mathsf{x}|\mathsf{y})$ is a functional of the explicitly parameterized prior $p(\mathsf{y})$ and the generating conditional $p(\mathsf{x}|\mathsf{y})$. Effectively, our parameterization of the encoder model is analogous to that of a conditionally trained discriminative model; however, in contrast to discriminative models, the lower-dimensional vectors y will in our case be hidden. Finally, we note that training encoder models by optimizing the likelihood would be meaningless, since the unknown representations y would marginalize out. On the other hand, training such models by maximizing the mutual information (1) will generally require approximations.

1.2 Linsker's *as-if Gaussian* Approximation

One way to tractably approximate $I(x, y)$ is by using the *as-if Gaussian* approximation of the joint encoder model $\mathcal{M}_I \stackrel{\text{def}}{=} p(x)p(y|x) \approx p_G(x, y) \sim \mathcal{N}(\boldsymbol{\mu}, \boldsymbol{\Sigma})$ [9]. The conditional entropy $H(x|y)$ may in this case be approximated by

$$H_G(x|y) \stackrel{\text{def}}{=} -\langle \log p_G(x|y) \rangle_{p_G(x,y)} = (1/2) \log(2\pi e)^{|x|} |\boldsymbol{\Sigma}_{x|y}|, \tag{2}$$

where $\boldsymbol{\Sigma}_{x|y}$ is the covariance of the Gaussian decoder $p_G(x|y)$ expressed from $p_G(x, y)$. The as-if Gaussian approximation of $I(x, y)$ is given by

$$I_G(x, y) \propto \log |\boldsymbol{\Sigma}_{xx}| - \log |\boldsymbol{\Sigma}_{xx} - \boldsymbol{\Sigma}_{xy} \boldsymbol{\Sigma}_{yy}^{-1} \boldsymbol{\Sigma}_{xy}^T|, \tag{3}$$

where $\boldsymbol{\Sigma}_{xx}$, $\boldsymbol{\Sigma}_{xy}$, and $\boldsymbol{\Sigma}_{yy}$ are the partitions of $\boldsymbol{\Sigma} \stackrel{\text{def}}{=} \langle [x\ y][x\ y]^T \rangle_{p(y|x)\tilde{p}(x)} - \langle [x\ y] \rangle_{p(y|x)\tilde{p}(x)} \langle [x\ y]^T \rangle_{p(y|x)\tilde{p}(x)}$, and we have assumed that $p(x) \equiv \tilde{p}(x)$ is the empirical distribution. Objective (3) needs to be maximized with respect to parameters of the encoder distribution $p(y|x)$. After training, the encoder may be used for generating lower-dimensional representations y for a given source x (note that the inference is usually much simpler than that in generative models).

2   A Simple Variational Lower Bound on $I(x, y)$

In [1] we discussed properties of a simple variational lower bound on the mutual information $I(x, y)$. The bound follows from non-negativity of the Kullback-Leibler divergence $KL(p(x|y)||q(x|y))$ between the exact posterior $p(x|y)$ and its variational approximation $q(x|y)$, leading to

$$I(x, y) \geq \tilde{I}(x, y) \stackrel{\text{def}}{=} H(x) + \langle \log q(x|y) \rangle_{p(x,y)}, \tag{4}$$

where $q(x|y)$ is an arbitrary distribution. Clearly, the bound is saturated for $q(x|y) \equiv p(x|y)$; however, in general this choice would lead to intractability of learning the optimal encoder $p(y|x)$.

Objective (4) explicitly includes both the encoder $p(y|x)$ (distribution of the lower-dimensional representations for a given source) and decoder $q(x|y)$ (reconstruction of the source from a given compressed representation). It is iteratively optimized for parameters of both distributions (the *IM* algorithm [1]), which is qualitatively similar to the variational expectation-maximizing algorithm for intractable generative models [10]. (Note, however, that optimization surfaces

defined by the objectives of the IM and the variational EM are quite differ-
ent [6]). The flexibility in the choice of the decoder $q(x|y)$ makes (4) particularly
computationally convenient. Indeed, we may avoid most of the computational
difficulties of optimizing $I(x, y)$ by constraining $q(x|y)$ to lie in a tractable fam-
ily (for example, $q(x|y)$ may be chosen to have a simple parametric form or a
sparse structure). Such constraints significantly simplify optimization of $\tilde{I}(x, y)$
for non-trivial stochastic encoders and provide a variational extension of the
Blahut-Arimoto algorithms [11] for channel capacity[1].

It is easy to show that by constraining the decoder as $q(x|y) \sim \mathcal{N}(Uy, \Sigma)$, op-
timization of the bound (4) reduces to maximization of Linsker's *as-if Gaussian*
criterion (3) (see [6] for details). Therefore, maximization of I_G may be seen
as a special case of the variational information-maximization approach for the
case when the decoder $q(x|y)$ is a linear Gaussian. Moreover, if for this case
$p(y|x) \sim \mathcal{N}\left(W(x - \langle x \rangle_{\tilde{p}(x)}), s^2 I\right)$, it is easy to show that the left singular vectors
of the optimal projection weights W^T correspond to the $|y|$-PCA solution on the
sample covariance $\langle xx^T \rangle_{\tilde{p}(x)} - \langle x \rangle_{\tilde{p}(x)} \langle x \rangle_{\tilde{p}(x)}^T$.

3   An Auxiliary Variational Bound

A principal conceptual difficulty of applying the bound (4) is in specifying a pow-
erful yet tractable variational decoder $q(x|y)$. Specifically, for isotropic Gaussian
channels, the linear Gaussian decoders mentioned above are fundamentally limited to producing PCA projections. Here we describe a richer family of bounds
on $I(x, y)$ which helps to overcome this limitation.

From (4) it is intuitive that we may obtain tighter bounds on $I(x, y)$ by in-
creasing representational power of the variational distributions $q(x|y)$. One way
to achieve this is to consider multi-modal decoders $q(x|y) = \langle q(x|y, z) \rangle_{q(z|y)}$, where
the introduced *auxiliary variables* z are effectively the unknown mixture states.
Clearly, this choice of the variational decoder has a structure of a constrained
multi-dimensional mixture-of-experts [14] model of a conditional distribution
(though in our case the lower-dimensional representations y are hidden). The
fully-coupled structure of the resulting variational distribution $q(x|y)$ qualita-
tively agrees with the structure of the exact posterior, as different dimensions of
the reconstructed vectors x are coupled through the auxiliary variables z. More-
over, for any interesting choice of the auxiliary space $\{z\}$, the decoder $q(x|y)$ will
typically be multi-modal, which agrees with the generally multi-modal form of
Bayesian decoders $p(x|y)$. We may therefore intuitively hope that this choice of
the variational posterior will generally result in tighter bounds on $I(x, y)$.

A possible disadvantage of mixture decoders $q(x|y)$ relates to the fact that
specifying the conditional mixing coefficients $q(z|y)$ in a principled manner may
be rather difficult. Moreover, if the auxiliary variables z are independent from

---

[1] Standard iterative approaches to maximizing $I(x, y)$ in encoder models require optimization of the cross-entropy $\langle \log p^{(old)}(y) \rangle_{p(y)}$ between two mixture distributions
$p(y)$ and $p^{(old)}(y)$ for parameters of the encoder $p(y|x)$ (see [11], [12], [13]), which is
rarely tractable in practice.

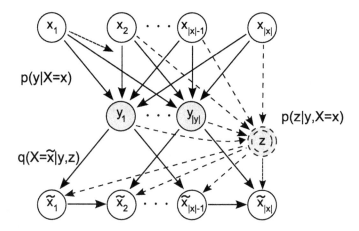

Fig. 2. A stochastic channel $p(y|x)$ with a structured mixture-type decoder $q(x|y)$. (The states of the reconstructed variables are denoted by $\tilde{x}$). The auxiliary vector z is *not* transmitted across the channel $p(y|x)$ and does not explicitly constrain $p(x,y)$. The dashed lines show the mappings to and from the auxiliary space. The auxiliary node is shown by the double circle.

the *original* source patterns x given the lower-dimensional encodings y, any noise in y will affect determining of the mixing states. Intuitively, this may have an overwhelmingly negative effect on decoding, causing relaxations in the bound on $I(x,y)$. We may therefore wish to reduce the effects which the noise of the stochastic projection $p(y|x)$ has on the specification of the decoder $q(x|y)$. One way to address this matter is by introducing an additional mapping $p(z|x,y)$ to the auxiliary variable space, which may be thought of as an additional variational parameter (see Fig. 2). Indeed, even when the channel is noisy, the conditional dependence of the auxiliary variables z on the unperturbed source patterns could result in an accurate detection of the states of the auxiliary variables. Note that the *auxiliary conditional* distribution $p(z|x,y)$ is defined in a way that does not affect the original noisy channel $p(y|x)$, as the channel would remain a marginal of the joint distribution of the original sources, codes, and auxiliary variables

$$p(x,y,z) = \tilde{p}(x)p(y|x)p(z|x,y).$$ (5)

The role of the auxiliary variables z in this context would be to capture global *features* of the transmitted sources, and use these features for choosing optimal decoder experts. Importantly, the auxiliary variables z are *not* transmitted across the channel. Their purpose here is to define a richer family of bounds on $I(x,y)$ which would generalize over objectives with simple constraints on variational decoders (such as linear Gaussians).

From the definition (5) and the chain rule for mutual information (e.g. [11]), we may express $I(y,x)$ as

$$I(y,x) = I(\{z,y\},x) - I(x,z|y),$$ (6)

where $I(\{z,y\}, x) \stackrel{\text{def}}{=} H(x) - H(x|z,y)$ is the amount of information that the features z and codes y jointly contain about the sources, and $I(x, z|y) \stackrel{\text{def}}{=} H(z|y) - H(z|x,y)$ is the conditional mutual information. Substituting the definitions into (6), we obtain a general expression of the mutual information $I(x, y)$ as a function of conditional entropies of the sources, codes, and auxiliary variables

$$I(y, x) = H(x) + H(z|x, y) - H(x|y, z) - H(z|y). \tag{7}$$

Then by analogy with (4) we obtain

$$I(y, x) \geq H(x) + H(z|x, y) + \langle \log q(x|y, z) \rangle_{p(x,y,z)} + \langle \log q(z|y) \rangle_{p(y,z)}. \tag{8}$$

Symbolically, (8) has a form vaguely reminiscent of the objectives optimized by *Information Bottleneck* (IB) methods [15]. However, the similarity is deceptive both conceptually and analytically, which is easy to see by comparing the optimization surfaces and the extrema. Additionally, we note that the auxiliary variational method is applicable to significantly more complex channels, provided that the variational distributions are appropriately constrained.

The mapping $p(z|x, y)$ to the feature space may be constrained so that the averages in (8) are tractable. For example, we may choose

$$p(z_j|x, y) = p(z_j|x) \propto \exp\{-(v_j^T x + b_j)\}, \tag{9}$$

where z_j is the jth state of a multinomial variable z. Analogously, we may constrain the variational decoders $q(x|y, z)$ and $q(z|y)$. In a specific case of a linear Gaussian channel $p(y|x) \sim \mathcal{N}(Wx, s^2 I)$, we may assume $q(x, z|y) \propto q(x|y, z)q(z)$ with $q(x|y, z_j) \sim \mathcal{N}(U_j y, S_j)$. Then objective (8) is optimized for the channel encoder, variational decoder, and the auxiliary conditional distributions, which is tractable for the considered parameterization. Effectively, we will still be learning a noisy linear projection, but for a different (mixture-type) variational decoder.

3.1 Learning Representations in the Augmented $\{y, z\}$-Space

Now suppose that the multinomial auxiliary variable z is actually observable at the receiver's end of the channel. Under this assumption, we may consider maximizing an alternative bound $\tilde{I}_H(x, \{y, z\}) \leq I(x, \{y, z\})$, defined by analogy with (4). (We will use the notation I_H to indicate that the channel $x \to \{y, z\}$ is generally heterogeneous; for example, z may be a generally unknown class label, while $y \in \mathbb{R}^{|y|}$ may define a lower-dimensional projection). This leads to a slight simplification of (8), which effectively reduces to

$$\tilde{I}_H(x, \{y, z\}) = H(x) + \langle \log q(x|y, z) \rangle_{\tilde{p}(x)p(y|x)p(z|x)}, \tag{10}$$

where the cross-entropic term is given by

$$\langle \log q(x|y, z) \rangle_{p(x,y,z)} = -\frac{1}{2M} \sum_{j=1}^{|z|} \sum_{i=1}^{M} p(z_j|x^{(i)}) \text{tr} \left\{ S_j^{-1} \left(d_j^{(i)} d_j^{(i)T} + s^2 U_j U_j^T \right) \right\}$$

$$- \frac{1}{2M} \sum_{j=1}^{|z|} \log |S_j| \sum_{i=1}^{M} p(z_j|x^{(i)}). \tag{11}$$

Here we ignored the irrelevant constants and defined

$$\mathsf{d}_j^{(i)} \overset{\text{def}}{=} \mathsf{x}^{(i)} - \mathsf{U}_j\mathsf{W}\mathsf{x}^{(i)} \in \mathbb{R}^{|\mathsf{x}|} \tag{12}$$

to be the distortion between the ith pattern and its reconstruction from a noise-less code at the mean of $q(\mathsf{x}|\mathsf{y}, z_j)$. From (10) – (12) it is easy to see that small values of the distortion terms $\mathsf{d}_j^{(i)}$ lead to improvements in the bound on $I(\mathsf{x}, \{\mathsf{y}, z\})$, which agrees with the intuition that the trained model should favour accurate reconstructions of the source patterns from their compressed representations.

Note that in the communication-theoretic interpretation of the considered heterogeneous channel, the auxiliary variables z will need to be communicated over the channel (*cf* bound (8)). Generally, this comes at a small increase in the communication cost, which in this case is $\sim O(\log|z|)$. For the model parameterization considered here, this corresponds to sending (or storing) an additional positive integer z, which would effectively index the decoder used at the reconstruction. Generally, the lower-dimensional representations of $\{\mathsf{x}\}$ will include not only the codes $\{\mathsf{y}\}$, but also the auxiliary labels z. Finally, we may note that unless $p(z|\mathsf{x})$ is strongly constrained, the mapping $\mathsf{x} \to z$ will typically tend to be nearly noiseless, as this would decrease $H(z|\mathsf{x})$ and maximize $I(\mathsf{x}, \{\mathsf{y}, z\})$.

4 Demonstrations

Here we demonstrate a few applications of the method to extracting optimal subspaces for the digits dataset. In all cases, we assumed that $|\mathsf{y}| < |\mathsf{x}|$. We also assumed that $p(\mathsf{x})$ is the empirical distribution.

4.1 Hand-Written Digits: Comparing the Bounds

In the first set of experiments, we compared optimal lower bounds on the mutual information $I(\mathsf{x}, \mathsf{y})$ obtained by maximizing the as-if Gaussian $I_G(\mathsf{x}, \mathsf{y})$ and the auxiliary variational $\tilde{I}(\mathsf{x}, \mathsf{y})$ objectives for hand-written digits (a part of the reduced MNIST [16] dataset). The dataset contained $M = 30$ gray-scaled instances of 14-by-14 digits 1, 2, and 8 (10 of each class), which were centered and normalized. The goal was to find a noisy projection of the $|\mathsf{x}| = 196$-dimensional training data into a $|\mathsf{y}| = 6$-dimensional space, so that the bounds $I_G(\mathsf{x}, \mathsf{y})$ and $\tilde{I}(\mathsf{x}, \mathsf{y})$ were maximized. We considered a linear Gaussian channel with an irreducible white noise, which in this case leads to the encoder distribution $p(\mathsf{y}|\mathsf{x}) \sim \mathcal{N}_\mathsf{y}\left(\mathsf{W}\mathsf{y}, s^2\mathsf{I}\right)$ with $\mathsf{W} \in \mathbb{R}^{6 \times 196}$. Our specific interest was in finding optimal *orthonormal* projections, so the weights were normalized to satisfy $\mathsf{W}\mathsf{W}^T = \mathsf{I}_{|\mathsf{y}|}$ (by considering the parameterization $\mathsf{W} = (\tilde{\mathsf{W}}\tilde{\mathsf{W}}^T)^{-1/2}\tilde{\mathsf{W}}$ with $\tilde{\mathsf{W}} \in \mathbb{R}^{|\mathsf{y}| \times |\mathsf{x}|}$). Effectively, this case corresponds to finding the most informative compressed representations of the source vectors for improving communication of the *non-Gaussian* data over a noisy Gaussian channel (by maximizing lower bounds on the channel capacity). Our specific interest here was to find whether we may indeed improve on Linsker's as-if Gaussian bound on the mutual information (with the optima given

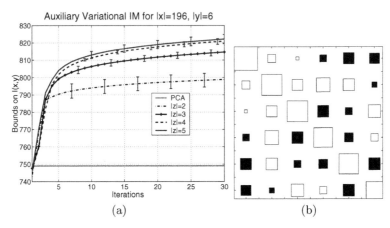

(a)   (b)

Fig. 3. Variational information maximization for noisy constrained dimensionality reduction. **(a):** *Top curves:* Average values of the variational auxiliary bounds $\tilde{I}(\mathsf{x}, \mathsf{y})$, obtained by the IM algorithm started at 10 random model initializations (shown for $|z| = 2, \ldots, 5$); *bottom line:* the *as-if* Gaussian $I_G(\mathsf{x}, \mathsf{y})$ bound (computed numerically). The results are shown for the digits data with $|\mathsf{x}| = 196$, $|\mathsf{y}| = 6$ for $M = 30$ patterns and $T = 30$ iterations of the IM. **(b):** Hinton diagram for $\mathsf{WW}_{pca}^T (\mathsf{WW}_{pca}^T)^T \in \mathbb{R}^{6 \times 6}$ for $|z| = 3$, $T = 30$. For orthonormal weights spanning identical subspaces, we would expect to see the identity matrix.

in this case by the PCA projection) by considering a richer family of auxiliary variational bounds with multi-modal mixture-type decoders.

Figure 3 shows typical changes in the auxiliary variational bound $\tilde{I}(\mathsf{x}, \mathsf{y})$ as a function of the IM's iterations T for $|z| \in \{2, \ldots, 5\}$ states of the discrete auxiliary variable. (On the plot, we ignored the irrelevant constants $H(\mathsf{x})$ identical for both $\tilde{I}(\mathsf{x}, \mathsf{y})$ and $I_G(\mathsf{x}, \mathsf{y})$, and interpolated $\tilde{I}(\mathsf{x}, \mathsf{y})$ for the consecutive iterations). The mappings were parameterized as described in Section 3, with the random initializations of the parameters v_j and b_j around zero, and the initial settings of the variational prior $q(z) = 1/|z|$. The encoder weights W were initialized at 6 normalized principal components $\mathsf{W}_{pca} \in \mathbb{R}^{6 \times 196}$ of the sample covariance $\langle \mathsf{xx}^T \rangle$, and the variance of the channel noise was fixed at $s^2 = 1$. For each choice of the auxiliary space dimension $|z|$, Figure 3 (a) shows the results averaged over 30 random initializations of the IM algorithm. As we see from the plot, the IM learning leads to a consistent improvement in the auxiliary variational bound, which (on average) varies from $\tilde{I}_0(\mathsf{x}, \mathsf{y}) \approx 745.7$ to $\tilde{I}_T(\mathsf{x}, \mathsf{y}) \approx 822.2$ at $T = 30$ for $|z| = 5$. Small variances in the obtained bounds ($\sigma_T \approx 2.6$ for $T = 30$, $|z| = 5$) indicate a stable increase in the information content independently of the model initializations. From Figure 3 (a) we can also observe a consistent improvement in the average $\tilde{I}(\mathsf{x}, \mathsf{y})$ with $|z|$, changing as $\tilde{I}_{10}(\mathsf{x}, \mathsf{y}) \approx 793.9$, ≈ 806.3, ≈ 811.2, and ≈ 812.9 for $|z| = 2, \ldots, 5$ after $T = 10$ IM's iterations. In comparison, the PCA projection weights W_{pca} result in $I_G(\mathsf{x}, \mathsf{y}) \approx 749.0$, which is visibly worse than the auxiliary bound with the optimized parameters, and is just a little better than $\tilde{I}(\mathsf{x}, \mathsf{y})$ computed at a random initialization of the variational decoder for $|z| \geq 2$.

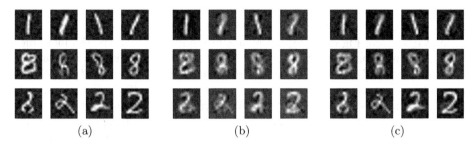

Fig. 4. Reconstructions of the source patterns from encoded representations. (a): A subset of the generic patterns used to generate the source vectors; (b): the corresponding reconstructions from 6 principal components; (c): the corresponding $\tilde{I}_H$-optimal reconstructions at $\langle \mathsf{x} \rangle_{q(\mathsf{x}|\mathsf{y},z)} = \mathsf{U}_z \mathsf{y}$ for the hybrid $\{\mathsf{y}, z\}$ representations ($|\mathsf{y}| = 6$, $|z| = 3$).

Importantly, we stress once again that the auxiliary variables z are not passed through the channel. In the specific case which we considered here, the auxiliary variables were used to define a more powerful family of variational bounds which we used to extract the $\tilde{I}$-optimal orthonormal subspace. The results are encouraging, as they show that for a specific constrained channel distribution we may indeed obtain more accurate lower bounds on the mutual information $I(\mathsf{x}, \mathsf{y})$ *without* communicating more data than in the conventional case. Specifically, for Gaussian channels with orthonormal projections to the code space, we do improve on simple *as-if Gaussian* solutions (leading to the PCA projections) by considering optimization of the auxiliary variational bounds (8).

As expected, we may also note that the $\tilde{I}$-optimal *encoder* weights W are in general different from rotations of W_{pca}. This is easy to see by computing $\mathsf{W}\mathsf{W}_{pca}^T(\mathsf{W}\mathsf{W}_{pca}^T)^T$, which in our case is visibly different from the identity matrix (see Fig. 3 (b) for $|\mathsf{y}| = 6$ and $|z| = 3$), which we would have expected to obtain otherwise. This indicates the intuitive result that by allowing a greater flexibility in the choice of the *variational decoder* distributions, the $\tilde{I}(\mathsf{x}, \mathsf{y})$-optimal constrained *encoders* become different from the optimal encoders of simpler models (such as PCA under the linear Gaussian assumption).

4.2 Hand-Written Digits: Reconstructions

Additionally, for the problem settings described in Sec. 4.1, we have computed reconstructions of the source patterns $\{\mathsf{x}\}$ from their noisy encoded representations. First, we generated source vectors by adding an isotropic Gaussian noise to the generic patterns (see Fig. 4 (a)), where the variance of the source noise was set as $s_s^2 = 0.5$. Then we computed noisy linear projections $\{\mathsf{y}\}$ of the source vectors by using the I_G- and the $\tilde{I}_H$- optimal encoder weights (in the latter case, we also computed the auxiliary label z by sampling from the learned $p(z|\mathsf{x})$). This stage corresponds to passing encoded representations over the noisy channels, where the noise variance for the Gaussian part of the channel was fixed at $s^2 = 1$. Finally, we have used the optimal *approximate* decoders to perform the reconstructions from $\{y\}$ (for I_G-optimal PCA projections) and $\{y, z\}$ (for $\tilde{I}_H$-optimal hybrid channels).

As we see from Figure 4 (b), (c), by a slight modification of the channel (due to encoding and communicating a multinomial variable z), we may achieve a visible improvement in the reconstruction of the sources by using the $\tilde{I}_H$- optimal projections[2]. The results are shown for $|y| = 6$, $|z| = 3$ after $T = 3$ iterations, and the reconstructions are computed at the analytical mean of the decoder's component $q(x|y, z)$ indexed by the auxiliary variable z. Even though the resulting hybrid channel may be difficult to justify from the communication viewpoint, the results suggest that maximization of the bound on $I(x, \{y, z\})$ provides a sensible way to reduce dimensionality of the sources for the purpose of reconstructing inherently noisy non-Gaussian patterns. Importantly, the variational decoder $q(z|x, y)$ which maximizes $\tilde{I}_H(x, \{y, z\})$ makes no recourse to $\tilde{p}(x)$. Therefore, just like in the PCA case, we do not need to store the training instances in order to perform an accurate reconstruction from noisy lower-dimensional projections. We note once again that the weights of the optimal encoder were chosen to satisfy the specific orthonormality constraint (though other kinds of constrained encoders may easily be considered). This contrasts with the exact approaches to training generative models, where encoding constraints may be more difficult to enforce.

5   Summary

Here we described an auxiliary variational approach to information maximization, and applied it to linear orthonormal dimensionality reduction in the presence of irreducible Gaussian noise. For this case we can show that the common *as-if* Gaussian [9] approximation of MI is in fact a suboptimal special case of our variational bound, which for isotropic linear Gaussian channels leads to the PCA solution. Importantly, this means that by using linear Gaussian variational decoders under the considered Gaussian channel, maximization of the generic lower bound (4) on MI cannot improve on the PCA projections. The situation changes if we consider a richer family of *variational auxiliary* lower bounds on $I(x, y)$ under the same encoding constraints. In particular, we showed that in the cases when the source distribution was non-Gaussian, we could significantly improve on the PCA projections by considering multi-modal variational decoders. This confirms the conceptually simple idea that by allowing a greater flexibility in the choice of variational decoders, we may get significant improvements over simple bounds at a limited increase in the computational cost. This result is also interesting from the communication-theoretic perspective, as it demonstrates a simple and computationally efficient way to produce better bounds on the capacity of communication channels without altering channel properties (e.g. without communicating more data across the channels). Finally, we discussed a simple

---

[2] Similar reconstructions could be obtained by maximizing the auxiliary bound $\tilde{I}(x, y)$ *without* communicating z. However, the approximate decoder for this case would be given as $q(x|y) = \sum_z q(x|y, z) \frac{\langle p(z|x)p(y|x)\rangle_{p(x)}}{\langle p(z|x)\rangle_{p(x)}}$, which requires knowing $p(x)$.

information-theoretic approach to constrained dimensionality reduction for hybrid representations $x \to \{y, z\}$, which may significantly improve reconstructions of the sources $\{x\}$ from their lower-dimensional representations $\{y\}$ at a small increase in the transmission cost.

It is potentially interesting to compare the variational information-maximizing framework with other approaches applicable to learning unknown under-complete representations of the data (such as generative models[3] and autoencoders). As we pointed out, there are important conceptual differences in the way we parameterize and train encoder and generative models. Specifically, by imposing explicit constraints on the mapping to the space of representations, our method is applicable for *constrained stochastic* dimensionality reduction. This may be particularly useful in engineering and neural systems, where such constraints may be physically or biologically motivated. It is also interesting to note that despite important conceptual differences, the special case of the auxiliary variational bound on $I(x, y)$ for a Gaussian channel and a multinomial auxiliary space $\{z\}$ has an interesting link to likelihood maximization for a mixture of factor-analysis-type models with the *uniform*, rather than Gaussian, factor distribution [6] (*cf* [8]).

It is also interesting to compare our framework with self-supervised training in semi-parametric models. The most common application of self-supervised models is dimensionality reduction in autoencoders $x \to y \to \tilde{x}$, where $x^{(m)} = \tilde{x}^{(m)}$ for all patterns m. Typically, it is presumed that $y = f(x)$, and the models are trained by minimizing a loss function (such as the squared loss). It is clear that for noiseless encoders, our bound (4) gives $const + \sum_m \log q\left(x^{(m)}|y = f(x^{(m)})\right)$, which has the interpretation of an autoencoder whose loss function is determined by q. Thus a squared loss function can be interpreted as an assumption that the data x can be reconstructed from *noiseless* codes y with Gaussian fluctuations. However, in some sense, the natural loss function (from the MI viewpoint) would not be the squared loss, but that which corresponds to the Bayesian decoder $q(x|y) = p(x|y)$, and more powerful models should strive to approximate this. Indeed, this is also the role of the auxiliary variables – effectively to make a loss function which is closer to the Bayes optimum. What is also interesting about our framework is that it holds in the case that the codes are stochastic, for which the autoencoder framework is more clumsy. Indeed, it also works when we have a (non-delta mixture) distribution $p(x)$, i.e. the method merges many interesting models in one framework.

References

1. Barber, D., Agakov, F.V.: The IM Algorithm: A Variational Approach to Information Maximization. In: NIPS, MIT Press (2003)
2. Linsker, R.: An Application of the Principle of Maximum Information Preservation to Linear Systems. In Touretzky, D., ed.: Advances in Neural Information Processing Systems. Volume 1., Morgan-Kaufmann (1989)

---

[3] It is well known that in a few special cases (e.g. for square ICA models) mutual information- and likelihood-maximization may lead to the same extrema [17]. However, little is understood about how the frameworks may relate in general.

3. Bell, A.J., Sejnowski, T.J.: An information-maximization approach to blind separation and blind deconvolution. Neural Computation **7**(6) (1995) 1129–1159
4. Brunel, N., Nadal, J.P.: Mutual Information, Fisher Information and Population Coding. Neural Computation **10** (1998) 1731–1757
5. Agakov, F.V., Barber, D.: Variational Information Maximization for Neural Coding. In: International Conference on Neural Information Processing, Springer (2004)
6. Agakov, F.V.: Variational Information Maximization in Stochastic Environments. PhD thesis, School of Informatics, University of Edinburgh (2005)
7. C. Bishop, M. Svensen and C. K. I. Williams: GTM: The Generative Topographic Mapping. Neural Computation **10**(1) (1998) 215–234
8. Tipping, M.E., Bishop, C.M.: Mixtures of Probabilistic Principal Component Analyzers. Neural Computation **11**(2) (1999) 443–482
9. Linsker, R.: Deriving Receptive Fields Using an Optimal Encoding Criterion. In Steven Hanson, Jack Cowan, L.G.e., ed.: Advances in Neural Information Processing Systems. Volume 5., Morgan-Kaufmann (1993)
10. Neal, R.M., Hinton, G.E.: A View of the EM Algorithm That Justifies Incremental, Sparse, and Other Variants. In Jordan, M., ed.: Learning in Graphical Models. Kluwer Academic (1998)
11. Cover, T.M., Thomas, J.A.: Elements of Information Theory. Wiley (1991)
12. Arimoto, S.: An Algorithm for computing the capacity of arbitrary discrete memoryless channels. IT **18** (1972)
13. Blahut, R.: Computation of channel capacity and rate-distortion functions. IT **18** (1972)
14. Jacobs, R.A., I.Jordan, M., Nowlan, S.J., Hinton, G.E.: Adaptive Mixtures of Local Experts. Neural Computation **3** (1991)
15. Tishby, N., Pereira, F.C., Bialek, W.: The information bottleneck method. In: Proc. of the 37-th Annual Allerton Conference on Communication, Control and Computing. (1999) 368–377
16. LeCun, Y., Cortes, C.: The MNIST database of handwritten digits (1998)
17. Cardoso, J.F.: Infomax and maximum likelihood for blind source separation. In: IEEE Signal Processing Letters. Volume 4. (1997)

Constructing Visual Models with a Latent Space Approach

Florent Monay, Pedro Quelhas, Daniel Gatica-Perez, and Jean-Marc Odobez

IDIAP Research Institute, 1920 Martigny, Switzerland
monay@idiap.ch, quelhas@idiap.ch, gatica@idiap.ch, odobez@idiap.ch

Abstract. We propose the use of latent space models applied to local invariant features for object classification. We investigate whether using latent space models enables to learn patterns of visual co-occurrence and if the learned visual models improve performance when less labeled data are available. We present and discuss results that support these hypotheses. Probabilistic Latent Semantic Analysis (PLSA) automatically identifies aspects from the data with semantic meaning, producing unsupervised soft clustering. The resulting compact representation retains sufficient discriminative information for accurate object classification, and improves the classification accuracy through the use of unlabeled data when less labeled training data are available. We perform experiments on a 7-class object database containing 1776 images.

1 Introduction

The bag-of-words model is one of the most common text document representations in information retrieval (IR), in which a fixed-size vector stores the occurrence of the words present in a document. Although the sequential relations between words are not preserved, this somewhat drastic simplification allows simple comparisons between documents, and produces good performance for document classification and retrieval [1].

A text corpus represented by a bag-of-words is an example of a collection of discrete data, for which a number of generative probabilistic models have been recently proposed [5, 2, 3, 6]. The models, able to capture co-occurrence information between word and documents, have shown promising results in text dimensionality reduction, feature extraction, and multi-faceted clustering. It is thus not a surprise that the interest in casting other data sources into this representation has increased; recent work in computer vision has shown that images and videos are suitable for a vector-space representation, both for visual tasks like object matching [14], object classification [17], and cross-media tasks like image auto-annotation [4, 9, 10].

We propose here to build visual models from images in a similar fashion, using a quantized version of local image descriptors, dubbed visterms [15, 14]. However, unlike related work, which has only used the basic bag-of-words [14, 17], we propose to use a probabilistic latent space model, namely Probabilistic Latent Semantic Analysis (PLSA) [5] to build visual models of objects.

C. Saunders et al. (Eds.): SLSFS 2005, LNCS 3940, pp. 115–126, 2006.

The different outcomes of this model are principally unsupervised feature extraction and automatic soft clustering of image datasets, that we recently studied in the context of *scene modeling* [12]. Independently, Sivic et al. compared two latent probabilistic models of discretized local descriptors to discover object categories in image collections [13]. The approach is closely related to what we propose in this paper and in [12], but fundamentally differs in the assumption of the latent structure of the data. In [13], the number of classes is assumed to be known a priori. In contrast we assume that an image is a mixture of latent aspects that are not necessarily limited to the number of object categories in the dataset. We consider latent aspect modeling not as a classification system in itself, but as a feature extraction process for supervised classification. We show (qualitatively and quantitatively) the benefits of our formulation, and its advantages over the simple vector-space formulation. Based on the results, we believe that the approach might be worth exploring in other vision areas.

The paper is organized as follows. Section 2 describes the specific probabilistic model. In Section 3 we discuss the image representation. Section 4 summarizes results regarding object clustering and classification, and Section 5 concludes the discussion.

2   Latent Structure Analysis

2.1   Bag-of-Words: Data Sparseness

The vector-space approach tends to produce high-dimensional sparse representations. Sparsity makes the match between similar documents difficult, especially if ambiguities exist in the vector-space. In the text case for example, different words might mean the same (synonymy) and a word can have several meanings (polysemy). This potentially leads to ambiguous data representations. In practice, such situation also occurs with visterms.

To overcome this problem, different probabilistic generative models [5, 2, 3, 6] have been proposed to learn the co-occurrence between elements in the vector-space in an unsupervised manner. The idea is to model a latent data structure from the co-occurrence of elements in a specific dataset, assuming their independence given a latent variable. The elements in the vector space are probabilistically linked through the latent *aspect* variable, which identifies a disambiguated lower-dimensional representation. One model that implements this concept is PLSA, which we briefly review in the following.

2.2   Probabilistic LSA

In a dataset of N_d documents represented as bag-of-words of size N_x, the PLSA model assumes that the joint probability of a document d_i and an element x_j from the vector-space is the marginalization of the N_z joint probabilities of d_i, x_j and an unobserved latent variable z_k called *aspect*:

$$P(x_j, d_i) = \sum_{k=1}^{N_z} P(x_j, z_k, d_i)$$

$$= P(d_i) \sum_{k=1}^{N_z} P(z_k \mid d_i) P(x_j \mid z_k). \tag{1}$$

Each document is a mixture of latent aspects, expressed by the conditional probability distribution of the latent aspects given each document d_i, $P(z \mid d_i)$. Each latent aspect z_k is defined by the conditional probability distribution $P(x \mid z_k)$ in Eq. 1. The parameters are estimated by the Expectation-Maximization (EM) procedure described in [5] which maximizes the likelihood of the observation pairs (x_j, d_i). The E-step estimates the probability of the aspect z_k given the element x_j in the document d_i (Eq. 2).

$$P(z_k \mid d_i, x_j) = \frac{P(x_j \mid z_k) P(z_k \mid d_i)}{\sum_{k=1}^{N_z} P(x_j \mid z_k) P(z_k \mid d_i)} \tag{2}$$

The M-step then derives the conditional probabilities $P(x \mid z_k)$ (Eq. 3) and $P(z \mid d_i)$ (Eq. 4) from the estimated conditional probabilities of aspects $P(z_k \mid d_i, x_j)$ and the frequency count of the element x_j in image d_i, $n(d_i, x_j)$.

$$P(x_j \mid z_k) = \frac{\sum_{i=1}^{N_d} n(d_i, x_j) P(z_k \mid d_i, x_j)}{\sum_{m=1}^{N_x} \sum_{i=1}^{N_d} n(d_i, x_m) P(z_k \mid d_i, x_m)} \tag{3}$$

$$P(z_k \mid d_i) = \frac{\sum_{j=1}^{N_x} n(d_i, x_j) P(z_k \mid d_i, x_j)}{n(d_i)} \tag{4}$$

To prevent over-fitting, the number of EM iterations is controlled by an early stopping criterion based on the validation data likelihood. Starting from a random initialization of the model parameters, the EM iterations are stopped when the criterion is reached. The corresponding latent aspect structure defined by the current conditional probability distributions $P(x \mid z_k)$ is saved. Derived from the vector-space representation, the inference of $P(z_k \mid d_i)$ can be seen as a feature extraction process and used for classification. It also allows to rank images with respect to a given latent aspect z_k, which illustrates the latent structure learned from the data.

3   Images as Bag-of-Visterms

Although global features such as global color histograms or global edge direction histograms are traditionally used to represent images, a promising recent research direction in computer vision is the use of local image descriptors. The combination of interest point detectors and invariant local descriptors has shown interesting capabilities of describing images and objects. We decided to use the Difference of Gaussians (DOG) point detector [7] and the Scale Invariant Feature Transform (SIFT) local descriptors [7], as proposed in recent studies [8].

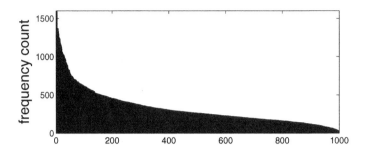

Fig. 1. Sorted document frequency counts of the quantized local image patches in the training set

The SIFT descriptors are local histograms of edge directions and therefore correspond to local image structures. Note that only gray-level information is used for this process.

The idea is to identify different types of local image patches occurring in the database to represent an image, similarly to the bag-of-words approach. As for the word ordering, the spatial information of the local descriptors is not encoded in the image representation. Those local image patches are obtained by a standard K-means quantization of the extracted SIFT descriptors in an image dataset, and are referred to as *visterms* (visual terms). As an analogy with text, the image representation is referred to as *bag-of-visterms* (BOV). We did not experiment the standard inverse document frequency (idf) weighting, but restricted our experiments to the unweighted BOV representation. As shown in Figure 3, the K-means quantization produces much more balanced document frequencies than what is encountered in text (Zipf's law), and the BOV representation therefore does not need to be compensated.

4   Image Modeling with PLSA

4.1   Data Description

To create the visterm vocabulary (K-means) we use a 3805-image dataset constructed from several sources. This includes 1002 building images (Zubud), 144 images of people and outdoors [11], 435 indoor images with people faces [17], 490 indoor images from the corel collection [16], 1516 city-landscape overlapped images from Corel [16] and 267 Internet photographic images. Interests points are identified on each image with the DOG point detector, a SIFT description of each point is computed and all SIFT descriptors are quantized with K-means to construct the visterms 'vocabulary'.

We propose to consider a 7-class dataset to evaluate classification [17]. The image classes are: faces (792), buildings (150), trees (150), cars (201), phones (216), bikes (125) and books (142), adding up to a total of 1776 images. The size of the images varies considerably: images can have between 10k and 1,2M pixels

aspect 3   aspect 17   aspect 8   aspect 10   aspect 5   aspect 7   aspect 12

Fig. 2. 10 top-ranked images with respect to $P(z_k \mid d_i)$ for seven selected aspects. Images are cropped for a convenient display. A full ranking is available at http://www.idiap.ch/~monay/PASCAL_LATENT/

while most image sizes are around 100-150k pixels. We resize all images to 100k pixels since the local invariant feature extraction process is highly dependent of the image size. This ensures that no class-dependent image size information is included in the representation. The dataset is split in 10 test sets, which allows ten evaluation runs with different training and test sets each time. We decided to use 1000 visterms to represent each image (size of the BOV).

4.2   Image Soft Clustering

The latent structure learned by PLSA can be illustrated by the top-ranked images in a dataset with respect to the posterior probabilities $P(z_k \mid d_i)$. Fig. 2 shows a ranking of seven out of 20 aspects identified by PLSA on the 7-class dataset described above. We selected $N_z = 20$ for a cleaner ranking visualization. From Fig. 2, we observe that aspects 3 and 17 seem closely related to face images. The first ten images ranked with respect to aspect 8 are all bike images, while top-ranked images for aspect 10 mostly contain phones. Buildings

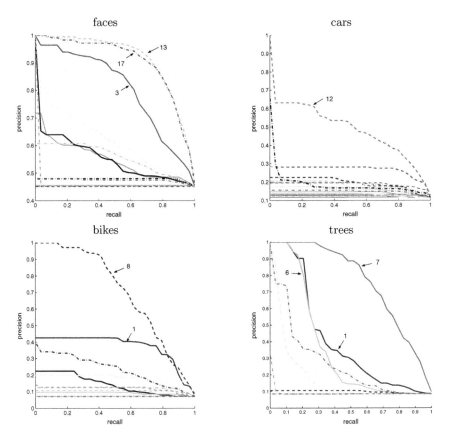

Fig. 3. Precision and recall curves for the 'face', 'car', 'bike' and 'tree' categories, according to an aspect-based unsupervised image ranking. The lowest precision values on the graph correspond to a random ranking.

are present in aspect 5, all images related to aspect 7 are tree images. Aspect 12 does not seem to be related to any specific object category.

To analyze the ranking in more details, the precision and recall curves for the retrieval of faces, cars, bikes, and trees are shown in Fig. 3. The top left graph shows that the homogeneous ranking holds on for more than 10 retrieved images in aspect 3 and 17, confirming the observations made from Fig. 2. We see that another aspect (13) is closely related to face images. The top right graph from Fig. 3 shows that aspect number 12 is related to car images if looking deeper in the ranking, what is not obvious from the observation of Fig. 2. Note however that the precision/recall values are not as high as for the faces case. The bottom left graph confirms that aspect 8 is linked to bike images, as well as aspect 1 even if less obvious. The bottom right graph shows that top-ranked images with respect to aspect 7 are mainly tree images. These results confirm that PLSA can capture class-related information in an unsupervised manner.

4.3   Images as Mixtures of Aspects

Our model explicitly considers an image as a mixture of latent aspects expressed by the $P(z \mid d)$ distributions learned from PLSA. The same latent structure with $N_z = 20$ aspects used for the aspect-based image ranking is considered. As illustrated by the aspect-based image ranking from Fig. 2, some identified aspects

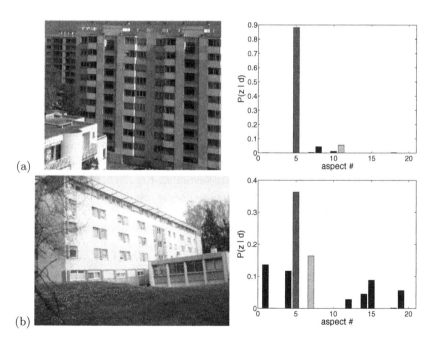

Fig. 4. Images and their corresponding aspect distribution $P(z \mid d)$ for $N_z = 20$. (a) is concentrated on aspect 5 (building), while (b) is a mixture of aspects 5 (building), 7 (tree) and aspect 1.

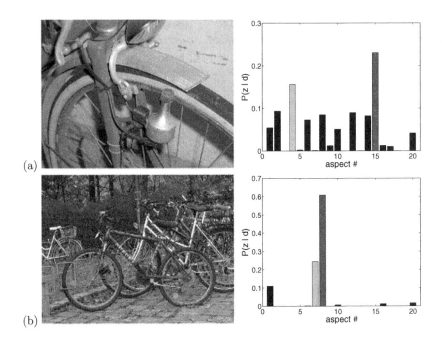

Fig. 5. Images and their corresponding aspect distribution $P(z \mid d)$ for $N_z = 20$. (a) is a mixture of different aspects, (b) is a mixture of aspect 8 (bikes) and 7 (trees).

relate to specific object categories. Within the dataset, different examples of aspect mixtures can be observed. In Fig. 4 (a) the aspect distribution is mainly concentrated on the aspect related to 'building' images. The image only contains building structures, therefore the aspect distribution seems coherent. On the contrary, the image from Fig. 4 (b) is composed of both 'building' and 'tree' -related structures. The corresponding aspect distribution interestingly reflects this image composition with the most probable aspects related to 'building' and 'tree'.

It is important to point out that there are cases when the aspect distribution does not clearly correspond to the image semantic. Fig. 5 (a) shows the close-up of a bike, but the aspect distribution is not concentrated on aspect 8, previously related to 'bike' images. The aspect distribution $P(z \mid d)$ rather describes the image as a mixture of several aspects with no specific dominance. This ambiguous aspect representation could derive from the fact that only a few examples of this type of close-up appear in the database. In Fig. 5 (b), the image is identified as a mixture of aspect 8 and 7, which perfectly reflects the image composition. Bikes are located in the image on a tree/vegetation background.

4.4   Feature Extraction

The PLSA model can be seen as a feature extraction or dimensionality reduction process: from the bag-of-visterms, a lower-dimensional aspect-based representation $P(z_k \mid d_i)$ is inferred using a previously learned PLSA model. Here we propose to compare the aspect-based and the bag-of-visterms representations on the 7-class supervised classification task. The PLSA model is trained on all non-test images each time and the resulting model is used to extract the aspect-based representation. To evaluate the quality of the feature extraction, we compare the classification based on the BOV representation with the aspect-based representation with the same classification setup: one Support Vector Machine (SVM) per class is trained with one class against all others.

Table 1. Confusion matrix for the 7-class object classification problem using the bag-of-visterms features, summed over 10 runs, and average classification error with the variance over ten runs indicated in brackets

	faces	buildings	trees	phones	cars	bikes	books	error
faces	772	2	7	3	3	2	3	2.5(0.04)
buildings	6	100	6	5	12	5	16	33.3(1.70)
trees	1	3	141	1	3	1	0	6.0(0.60)
phones	14	0	0	187	6	2	7	13.4(1.20)
cars	18	1	2	12	162	3	3	19.4(1.46)
bikes	0	3	3	1	2	116	0	7.2(0.38)
books	13	8	0	9	9	1	102	28.2(1.86)

Table 2. Confusion matrix for the 7-class object classification problem using PLSA with $N_z = 60$ aspects as a feature extraction process, summed over 10 runs, and average classification error with the variance over ten runs indicated in brackets

	faces	buildings	trees	phones	cars	bikes	books	error
faces	772	2	5	1	10	1	1	2.5(0.02)
buildings	2	113	3	3	18	5	6	24.6(1.40)
trees	3	3	140	0	2	2	0	6.7(0.40)
phones	9	5	0	166	23	2	11	23.1(0.60)
cars	14	5	0	3	172	4	3	14.4(0.67)
bikes	0	3	4	0	4	113	1	9.6(0.69)
books	7	13	0	6	14	0	102	28.2(1.54)

Table 1 and Table 2 show the confusion matrix for the BOV and the PLSA-based classification with $N_z = 60$ aspects. The last column is the per class error. We see that the classification performance greatly depends on the object class for both the BOV and the PLSA representations. These differences are caused by diverse factors. For instance 'trees' is a well defined class that is dominated by high frequency texture visterms, and therefore does not get confused with other classes. Similarly, most 'face' images have an homogeneous background and consistent layout which will not create ambiguities with other classes in the BOV representation. This explains the good performance of these two categories.

On the contrary, 'car' images present a large variability in appearance within the database. Front, side and rear car views on different types of background can

Table 3. Comparison between the bag-of-visterms (BOV) and the PLSA-based representation (PLSA) for classification with an SVM classifier trained with progressively less training data on the 7-class problem. The number in brackets is the variance over the different data splits.

Method	90%	50%	10%	5%
PLSA ($N_z = 60$)	11.1(1.6)	12.5(1.5)	18.1(2.7)	21.7(1.7)
BOV	11.1(2.0)	13.5(2.0)	21.8(3.6)	26.7(2.8)

be found, what makes it a highly complex category for object classification, generating an important confusion with other classes. 'Phones', 'books' and 'buildings' are therefore confused with 'cars' in both the BOV and the PLSA case. The 'bike' class is well classified despite a variability in appearance comparable to the 'car' images, because the bike structure generates a discriminative BOV representation.

Table 3 summarizes the whole set of experiments when we gradually train the SVM classifiers with less training data. If using all the training data (90% of all data) for feature extraction and classification, BOV and PLSA achieve a similar total error score. This proves that while achieving a dimensionality reduction from 1000 visterms to $N_z = 60$ aspects, PLSA keeps sufficient discriminative information for the classification task.

The case in which PLSA is trained on all the training data, while the SVMs are trained on a reduced data portion of it, it corresponds to a partially labeled data problem. Being completely unsupervised, the PLSA approach can take advantage of any unlabeled data and build the aspect-based representation from it. This advantage with respect to supervised strategies is shown in Table 3 for 50%, 10% and 5% training data. Here the comparison between BOV and PLSA is done for the same reduced number of labeled images to train the SVM classifiers, while the PLSA model is still trained on the full 90% training data. The total classification errors show that the features extracted by PLSA outperform the raw BOV representations for the same amount of labeled data. Note also that the variance over the splits is smaller, which suggests that the model is more stable given the reduced dimensionality.

5   Conclusion

For an object classification task, we showed that using PLSA on a bag-of-visterms representation (BOV) produces a compact, discriminative representation of the data, outperforming the standard BOV approach in the case of small amount of training data. Also, we showed that PLSA can capture semantic meaning in the BOV representation allowing for both unsupervised ranking of object images and description of images as a mixture of aspects. These results motivate further investigation of this and other latent space approaches for task related to object recognition.

Acknowledgments

This work was also funded by the European project "CARTER: Classification of visual Scenes using Affine invariant Regions and TExt Retrieval methods" part of "PASCAL: Pattern Analysis, Statistical Modeling and Computational Learning", through the Swiss Federal Office for Education and Science (OFES).

We thank the Xerox Research Center Europe (XRCE) and the University of Graz for collecting the object images and making the database available in the context of the Learning for Adaptable Visual Assistant (LAVA) project.

The authors acknowledge financial support provided by the Swiss National Center of Competence in Research (NCCR) on Interactive Multimodal Information Management (IM)2. The NCCR is managed by the Swiss National Science Foundation on behalf of the Federal Authorities.

References

1. R. Baeza-Yates and B. Ribeiro-Neto. *Modern Information Retrieval*. ACM Press, 1999.
2. D. Blei, Y. Andrew, and M. Jordan. Latent dirichlet allocation. *Journal of Machine Learning Research*, 3:993–1020, 2003.
3. W Buntine. Variational extensions to em and multinomial pca. In *Proc. of Europ. Conf. on Machine Learning*, Helsinki, Aug. 2002.
4. P. Duygulu, K. Barnard, N. Freitas, and D. Forsyth. Object recognition as machine translation: Learning a lexicon for a fixed image vocabulary. In *Proc. of IEEE Europ. Conf. on Computer Vision*, Copenhagen, Jun. 2002.
5. T. Hofmann. Unsupervised learning by probabilistic latent semantic analysis. *Machine Learning*, 42:177–196, 2001.
6. M. Keller and S. Bengio. Theme topic mixture model: A graphical model for document representation. *IDIAP Research Report, IDIAP-RR-04-05*, January 2004.
7. D. G. Lowe. Distinctive image features from scale-invariant keypoints. *International Journal of Computer Vision*, 60:91–110, 2003.
8. K. Mikolajczyk and C. Schmid. A performance evaluation of local descriptors. In *Proc. of IEEE Conf. on Computer Vision and Pattern Recognition*, Madison, Jun. 2003.
9. F. Monay and D. Gatica-Perez. On image auto-annotation with latent space models. In *Proc. of ACM Int. Conf. on Multimedia*, Berkeley, Nov. 2003.
10. F. Monay and D. Gatica-Perez. PLSA-based image auto-annotation: Constraining the latent space. In *Proc. ACM Int. Conf. on Multimedia*, New York, Oct. 2004.
11. A. Opelt, M. Fussenegger, A. Pinz, and P. Auer. Weak hypotheses and boosting for generic object detection and recognition. In *Proc. of IEEE Europ. Conf. on Computer Vision*, Prague, May 2004.
12. P. Quelhas, F. Monay, J.-M. Odobez, D. Gatica-Perez, T. Tuytelaars, and L. V. Gool. Modeling scenes with local descriptors and latent aspects. In *Proc. of IEEE Int. Conf. on Computer Vision*, Beijing, Oct. 2005.
13. J. Sivic, B. C. Russell, A. A. Efros, A. Zisserman, and W. T. Freeman. Discovering object categories in image collections. Technical report, Dept. of Engineering Science, University of Oxford, 2005.

14. J. Sivic and A. Zisserman. Video google: A text retrieval approach to object matching in videos. In *Proc. of IEEE Int. Conf. on Computer Vision*, Nice, Oct. 2003.

15. T. Tuytelaars and L. Van Gool. Content-based image retrieval based on local affinely invariant regions. In *Proc. of Visual99*, Amsterdam, Jun. 1999.

16. A. Vailaya, M. Figueiredo, A. Jain, and H.J. Zhang. Image classification for content-based indexing. *IEEE Trans. on Image Processing*, 10:117–130, 2001.

17. J. Willamowski, D. Arregui, G. Csurka, C. R. Dance, and L. Fan. Categorizing nine visual classes using local appearance descriptors. In *Proc. of ICPR Workshop on Learning for Adaptable Visal Systems*, Cambridge, Aug. 2004.

Is Feature Selection Still Necessary?

Amir Navot[2], Ran Gilad-Bachrach[1], Yiftah Navot, and Naftali Tishby[1,2]

[1] School of Computer Science and Engineering and
[2] Interdisciplinary Center for Neural Computation
The Hebrew University, Jerusalem, Israel
{anavot, ranb, tishby}@cs.huji.ac.il, yiftah@sarin.com

Abstract. Feature selection is usually motivated by improved computational complexity, economy and problem understanding, but it can also improve classification accuracy in many cases. In this paper we investigate the relationship between the optimal number of features and the training set size. We present a new and simple analysis of the well-studied two-Gaussian setting. We explicitly find the optimal number of features as a function of the training set size for a few special cases and show that accuracy declines dramatically by adding too many features. Then we show empirically that *Support Vector Machine* (SVM), that was designed to work in the presence of a large number of features produces the same qualitative result for these examples. This suggests that good feature selection is still an important component in accurate classification.

1 Introduction

Feature selection is the task of choosing a small subset out of a given set of features that captures the relevant properties of the data. In the context of supervised classification problems, relevance is determined by the assigned labels on the training data. The main advantages of feature selection are: reduced computational complexity, economy (as it saves the cost of measuring irrelevant features), insight into the problem at hand and improved classification accuracy. In this work we focus on the latter issue. It is well known that the presence of many irrelevant features can reduce classification accuracy. Many algorithms have been suggested for the task of feature selection, e.g. *Infogain* [1], *Relief* [2], *Focus* [3], selection using *Markov Blanket* [4] and *Margin Based* [5]. See [6] for a comprehensive discussion of feature selection methodologies.

This work looks at choosing the optimal subset of features in terms of classification accuracy. Obviously, if the true statistical model is known, or if the sample is unlimited, any additional feature can only improve accuracy. However, when the training set is finite, additional features can degrade the performance of many classifiers, even when all the features are statistically independent and carry information on the label. This phenomenon is sometimes called "the peaking phenomenon" and was already demonstrated more than three decades ago by [7, 8, 9] and in other works (see references there) on the classification problem of two Gaussian-distributed classes with equal covariance matrices (LDA).

C. Saunders et al. (Eds.): SLSFS 2005, LNCS 3940, pp. 127–138, 2006.

Recently, [10] analyzed this phenomenon for the case where the covariance matrices are different (QDA), however, this analysis is limited to the case where all the features have equal contributions. On the other hand [11] showed that, in the Bayesian setting, the optimal Bayes classifier can only benefit from using additional features. However, using the optimal Bayes classifier is usually not practical due to its computational cost and the fact that the true *prior* over the classifiers is not known. In their discussion, [11] raised the problem of designing classification algorithms which are computationally efficient and robust with respect to the feature space. Now, three decades later, it is worth inquiring whether today's state-of-the-art classifiers, such as *Support Vector Machine* (SVM) [12], achieve this goal.

In this work we re-visit the two-Gaussian classification problem, and concentrate on a simple setting of two spherical Gaussians. We present a new simple analysis of the optimal number of features as a function of the training set size. We consider the maximum likelihood estimation as the underlying classification rule. We analyze its error as function of the number of features and number of training instances, and show that while the error may be as bad as chance when using too many features, it approchs to the optimal error if we chose the number of features wisely. We also explicitly find the optimal number of features as a function of the training set size for a few specific examples. We test SVM empirically in this setting and show that its performance matches the predictions of our analysis. This suggests that feature selection is still a crucial component in designing an accurate classifier, even when modern discriminative classifiers are used, and even if computational constraints or measuring costs are not an issue.

The remainder of the paper is organized as follows: we describe the problem setting in section 2. Our analysis is presented in section 3. Next, we explicitly find the optimal number of features for a few examples in section 4. The results of applying SVM are presented in section 5.

2   Problem Setting and Notation

First, let us introduce some notation. Vectors in R^N are denoted by bold face lower case letter (e.g. x, μ) and the j'th coordinate of a vector x is denoted by x_j. We denote the restriction of a vector x to the first n coordinates by x_n.

Assume that we have two classes in R^N, labeled $+1$ and -1. The distribution of the points in the positive class is $Normal\,(\mu, \Sigma = I)$ and the distribution of the points in the negative class is $Normal\,(-\mu, \Sigma = I)$, where $\mu \in R^N$, and I is the $N \times N$ unit matrix. To simplify notation we assume, without loss of generality, that the coordinates of μ are ordered in descending order of their absolute value and that $\mu_1 \neq 0$; thus if we choose to use only $n < N$ features, the best choice would be the first n coordinates. The optimal classifier, i.e., the one that achieves the maximal accuracy, is $h\,(x) = \text{sign}\,(\mu \cdot x)$. If we are restricted to using only the first n features, the optimal classifier is $h\,(x, n) = \text{sign}\,(\mu_n \cdot x_n)$.

In order to analyze this setting we have to consider a specific way to estimate $\boldsymbol{\mu}$ from a training sample $S^m = \left\{\boldsymbol{x}^i, y^i\right\}_{i=1}^m$, where $\boldsymbol{x}^i \in R^N$ and $y^i \in \{+1, -1\}$ is the label of $\boldsymbol{x}^i$. We consider the maximum likelihood estimator of $\boldsymbol{\mu}$:

$$\hat{\boldsymbol{\mu}} = \hat{\boldsymbol{\mu}}\left(S^m\right) = \frac{1}{m}\sum_{i=1}^m y^i \boldsymbol{x}^i$$

Thus the estimated classifier is $\hat{h}\left(\boldsymbol{x}\right) = \mathrm{sign}\left(\hat{\boldsymbol{\mu}} \cdot \boldsymbol{x}\right)$. For a given $\hat{\boldsymbol{\mu}}$ and number of features n, we look on the generalization error of this classifier:

$$\mathrm{error}\left(\hat{\boldsymbol{\mu}}, n\right) = P\left(\mathrm{sign}\left(\hat{\boldsymbol{\mu}}_n \cdot \boldsymbol{x}_n\right) \neq y\right)$$

where y is the true label of $\boldsymbol{x}$. This error depends on the training set. Thus, for a given training set size m, we are interested in the average error over all the possible choices of a sample of size m:

$$\mathrm{error}\left(m, n\right) = E_{S^m}\mathrm{error}\left(\hat{\boldsymbol{\mu}}\left(S_m\right), n\right) \tag{1}$$

We look for the optimal number of features, i.e. the value of n that minimizes this error:

$$n_{opt} = \arg\min_n \mathrm{error}\left(m, n\right)$$

3   Analysis

For a given $\hat{\boldsymbol{\mu}}$, and n the dot product $\hat{\boldsymbol{\mu}}_n \cdot \boldsymbol{x}_n$ is a *Normal* random variable on its own and therefore the generalization error can be explicitly written as (using the symmetry of this setting):

$$\mathrm{error}\left(\hat{\boldsymbol{\mu}}, n\right) = P\left(\hat{\boldsymbol{\mu}}_n \cdot \boldsymbol{x}_n < 0 | + 1\right) = \Phi\left(-\frac{E_{\boldsymbol{x}}\left(\hat{\boldsymbol{\mu}}_n \cdot \boldsymbol{x}_n\right)}{\sqrt{V_{\boldsymbol{x}}\left(\hat{\boldsymbol{\mu}}_n \cdot \boldsymbol{x}_n\right)}}\right) \tag{2}$$

here Φ is the Gaussian cumulative density function: $\Phi\left(a\right) = \frac{1}{\sqrt{2\pi}}\int_{-\infty}^a e^{-\frac{1}{2}z^2}dz$. We denote by $E_{\boldsymbol{x}}$ and $V_{\boldsymbol{x}}$ expectation and variance with respect to the true distribution of $\boldsymbol{x}$.

For a given number of features n, and a given sample S, we have

$$E_{\boldsymbol{x}}\left(\hat{\boldsymbol{\mu}}_n \cdot \boldsymbol{x}_n\right) = \hat{\boldsymbol{\mu}}_n \cdot \boldsymbol{\mu}_n = \sum_{j=1}^n \hat{\mu}_j \mu_j$$

and

$$V_{\boldsymbol{x}}\left(\hat{\boldsymbol{\mu}}_n \cdot \boldsymbol{x}_n\right) = \sum_{j=1}^n \hat{\mu}_j^2 V_{\boldsymbol{x}}\left(x_j\right) = \sum_{j=1}^n \hat{\mu}_j^2$$

substituting in equation (2) we get:

$$\mathrm{error}\left(\hat{\boldsymbol{\mu}}, n\right) = \Phi\left(-\frac{\sum_{j=1}^n \hat{\mu}_j \mu_j}{\sqrt{\sum_{j=1}^n \hat{\mu}_j^2}}\right) \tag{3}$$

Now, for a given training set size m, We want to find n that minimizes the average error term E_{S^m} (error $(\hat{\mu}, n)$), but instead we look for n that minimizes an approximation of the average error:

$$n_{opt} = \arg\min_n \Phi \left(-\frac{E_{S^m}\left(\sum_{j=1}^n \hat{\mu}_j \mu_j\right)}{\sqrt{E_{S^m}\left(\sum_{j=1}^n \hat{\mu}_j^2\right)}} \right) \qquad (4)$$

We first have to justify why the above term approximates the average error. We look at the variance of the relevant terms (the numerator and the term in the square root in the denominator). $\hat{\mu}_j$ is a Normal random variable with expectation μ_j and variance $1/m$, thus

$$V_{S^m}\left(\sum_{j=1}^n \hat{\mu}_j \mu_j\right) = \sum_{j=1}^n \mu_j^2 V_{S^m}(\hat{\mu}_j) = \frac{1}{m}\sum_{j=1}^n \mu_j^2 \xrightarrow[m\to\infty]{} 0 \qquad (5)$$

$$V_{S^m}\left(\sum_{j=1}^n \hat{\mu}_j^2\right) = \frac{2n}{m^2} + \frac{4}{m}\sum_{\alpha=1}^n \mu_j^2 \xrightarrow[m\to\infty]{} 0 \qquad (6)$$

where in the last equality we used the fact that if $Z \sim Normal(\mu, \sigma^2)$, then $V(Z^2) = 2\sigma^4 + 4\sigma^2\mu^2$. We also note that:

$$E_{S^m}\left(\sum_{j=1}^n \hat{\mu}_j^2\right) = \sum_{j=1}^n \left(V_{S^m}(\hat{\mu}_j) + E(\hat{\mu}_j)^2\right)$$

$$= \sum_{j=1}^n \left(\mu_j^2 + \frac{1}{m}\right) \xrightarrow[m\to\infty]{} \sum_{j=1}^n \mu_j^2 > 0$$

therefore the denominator is not zero for any value of m (including the limit $m \to \infty$). Combining this with (5) and (6) and recalling that the derivative of Φ is bounded by 1, we conclude that, at least for a large enough m, it is a good approximation to move the expectation inside the error term. Figure 1 shows numerically that for specific choices of μ, moving the expectation inside the error term is indeed justified.

Now we turn to finding the n that minimizes (4), as function of the training set size m . This is equivalent to finding n that maximizes

$$f(n,m) = \frac{E_{S^m}\left(\sum_{j=1}^n \hat{\mu}_j \mu_j\right)}{\sqrt{E_{S^m}\left(\sum_{j=1}^n \hat{\mu}_j^2\right)}} = \frac{\sum_{j=1}^n \mu_j^2}{\sqrt{\sum_{j=1}^n \left(\mu_j^2 + \frac{1}{m}\right)}} = \frac{\sum_{j=1}^n \mu_j^2}{\sqrt{\frac{n}{m} + \sum_{j=1}^n \mu_j^2}} \qquad (7)$$

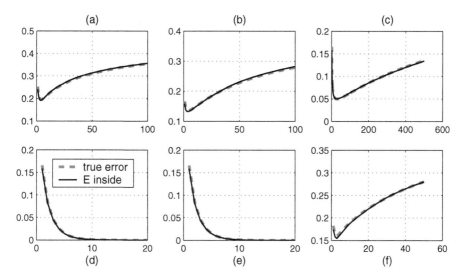

Fig. 1. Numerical justification for using equation 4. The true and approximate error ("E-inside") as function of the number of features used for different choices of μ: (a) $\mu_j = 1/\sqrt{2^j}$, (b) $\mu_j = 1/j$, (c) $\mu_j = 1/\sqrt{j}$, (d) $\mu_j = 1$, (e) $\mu_j = rand$ (sorted) and (f) $\mu_j = rand/j$ (sorted). The training set size here is $m = 16$. The true error was estimated by averaging over 200 repeats. The approximation with the expectation inside error term is very close to the actual error term (1), even for small training set ($m = 16$).

3.1   Observations on the Optimal Number of Features

First, we can see that, for any finite n, when $m \to \infty$, $f(n, m)$ reduces to $\sqrt{\sum_{j=1}^{n} \mu_j^2}$ and thus, as expected, using all the features maximizes it. It is also clear that adding a completely non-informative feature μ_j ($\mu_j = 0$) will decrease $f(n, m)$. We can also formulate a sufficient condition for the situation where using too many features is harmful, and thus feature selection can improve the accuracy dramatically:

Statement 1. *For the above setting, if the partial sum series $s_n = \sum_{j=1}^{n} \mu_j^2 < \infty$ then for any finite m the error of the ML classifier approaches to $1/2$ when $n \to \infty$ and there is $n_0 = n_0(m) < \infty$ such that selecting the first n_0 features is superior to selecting k features for any $k > n_0$.*

Proof. Denote $\lim_{n \to \infty} s_n = s < \infty$, then the numerator of (7) approaches s, while the denominator approaches ∞, thus $f(n, m) \xrightarrow[n \to \infty]{} 0$ and the error $\Phi(-f(n, m)) \to 1/2$. On the other hand $f(n, m) > 0$ for any finite n, thus there exists n_0 such that $f(n_0, m) > f(k, m)$ for any $k > n_0$.

Note that it is not possible to replace the condition in the above statement by $\mu_j \xrightarrow[j \to \infty]{} 0$ (see example 4). The following consistency statement gives a sufficient condition on the number of features that ensure asymptotic optimal error:

Statement 2. *For the above setting, if we use a number of features $n = n(m)$ that satisfies (1) $n \xrightarrow[m \to \infty]{} \infty$ and (2) $\frac{n}{m} \xrightarrow[m \to \infty]{} 0$ then the error of the ML estimator approaches to the optimal possible error (i.e. the error when μ is known and we use all the features) when $m \to \infty$. Additionally, if $\sum_{j=1}^{n} \mu_j \xrightarrow[n \to \infty]{} \infty$, condition (2) above can be replaced with $\frac{n}{m} \xrightarrow[m \to \infty]{} c$, where c is any finite constant.*

Proof. Recalling that the optimal possible error is given by $\Phi\left(-\sqrt{\sum_{j=1}^{\infty} \mu_j}\right)$, the statement follows directly from equation 7.

Corollary 1. *Using the optimal number of features ensure consistency (in the sense of the above statement).*

Note as well that the effect of adding a feature depends not only on its value, but also on the current value of the numerator and the denominator. In other words, the decision whether to add a feature depends on the properties of the features we have added so far. This may be surprising, as the features here are statistically independent. This apparent dependency comes intuitively from the signal-to-noise ratio of the new feature to the existing ones.

Another observation is that if all the features are equal, i.e. $\mu_j = c$ where c is a constant,

$$f(n, m) = \frac{c^2}{\sqrt{\frac{1}{m} + c^2}} \sqrt{n}$$

and thus using all the features is always optimal. In this respect our situation is different from the one analyzed by [7]. They considered Anderson's W classification statistic [13] for the setting of two Gaussians with same covariance matrix, but both the mean and the covariance were not known. For this setting they show that when all the features have equal contributions, it is optimal to use $m - 1$ features (where m is the number of training examples).

4   Specific Choices of μ

Now we find the optimal number of features for a few specific choices of μ.

Example 1. Let $\mu_j = \frac{1}{\sqrt{2^j}}$, i.e., $\mu = \left(\frac{1}{\sqrt{2}}, \frac{1}{\sqrt{4}}, \frac{1}{\sqrt{8}}, \ldots, \frac{1}{\sqrt{2^N}}\right)$, thus $\|\mu\| \xrightarrow[n \to \infty]{} 1$. An illustration of the density functions for this case is given in figure 2(a). Substituting this μ in (7), we obtain:

$$f(n, m) = \frac{1 - \frac{1}{2} 2^{-n}}{\sqrt{\frac{n}{m} + 1 - \frac{1}{2} 2^{-n}}}$$

Taking the derivative with respect to n and equating to zero we get[1] :

$$-\frac{1}{4}2^{-2n}\ln 2 + 2^{-n}\ln 2\left(1 + \frac{n}{m} + \frac{1}{2m}\right) - \frac{1}{m} = 0$$

Assuming n is large, we ignore the first term and get:

$$m = \frac{1}{\ln 2}2^n - n - \frac{1}{2\ln 2}$$

Ignoring the last two lower order terms, we have:

$$\frac{1}{\sqrt{m}} \cong \frac{\ln 2}{\sqrt{2^n}} = (\ln 2)\,\mu_n$$

This makes sense as $1/\sqrt{m}$ is the standard deviation of μ_j, so the above equation says that we only want to take features with a mean larger than the standard deviation. However, we should note that this is true **only in order of magnitude**. No meter how small the first feature is, it is worth to take it. Thus, we have no hope to be able to find optimal criterion of the form: take feature j only if $\mu_j > f\left(\sqrt{m}\right)$.

Example 2. Let $\mu_j = 1/j$. An illustration of the density functions for this case is given in figure 2(b). Since $\sum_{j=1}^{\infty}\left(\frac{1}{j}\right)^2 = \frac{\pi^2}{6}$,

$$\sum_{j=1}^{n}\left(\frac{1}{j}\right)^2 > \frac{\pi^2}{6} - \int_{n}^{\infty}\frac{1}{x^2}dx = \frac{\pi^2}{6} - \frac{1}{n}$$

$$\sum_{j=1}^{n}\left(\frac{1}{j}\right)^2 < \frac{\pi^2}{6} - \int_{n+1}^{\infty}\frac{1}{x^2}dx = \frac{\pi^2}{6} - \frac{1}{n+1}$$

thus the lower bound $\frac{\pi^2}{6} - \frac{1}{n}$ approximates the finite sum up to $\frac{1}{n} - \frac{1}{n+1} = \frac{1}{n(n+1)}$. Substituting this lower bound in (7), we obtain:

$$f(n,m) \cong \frac{\frac{\pi^2}{6} - \frac{1}{n}}{\sqrt{\frac{n}{m} + \frac{\pi^2}{6} - \frac{1}{n}}}$$

taking the derivative with respect to n and equating to zero yields $m \cong \frac{n^3\pi^2 - 18n^2}{n\pi^2 - 6}$ and thus for a large n we obtain the power law $m \cong n^2$, or $n \cong \sqrt{m} = m^{\frac{1}{2}}$, and again we have $\frac{1}{\sqrt{m}} \cong \frac{1}{n} = \mu_n$.

---

[1] The variable n is an integer, but we can consider $f(n,m)$ to be defined for any real value.

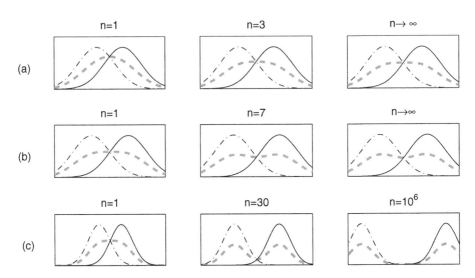

Fig. 2. Illustration of the separation between classes for different number of features (n), for different choices of $\boldsymbol{\mu}$. The projection on the prefix of $\boldsymbol{\mu}$ of the density function of each class and the combined density (in gray) are shown. (a) for example 1, (b) for example 2 and (c) for example 3.

Example 3. Let $\mu_j = \frac{1}{\sqrt{j}}$, i.e $\boldsymbol{\mu} = \left(1, \frac{1}{\sqrt{2}}, \dots \frac{1}{\sqrt{N}}\right)$. An illustration of the separation between classes for a few choices of n is given in figure 2(c). Substituting $\sum_{j=1}^{n} \mu_j^2 \cong \log n$ in (7) we get:

$$f(n, m) = \frac{\log n}{\sqrt{\frac{n}{m} + \log n}} \tag{8}$$

taking the derivative with respect to n and equating to zero, we obtain:

$$m \cong \frac{n (\log n - 2)}{\log n}$$

thus for a large n we have $m \cong n$, and once again we have $\frac{1}{\sqrt{m}} \cong \frac{1}{\sqrt{n}} = \mu_n$.

This example was already analyzed by Trunk ([8]) who showed that for any finite number of training examples, the error approaches one half when the number of features approaches ∞. Here we get this results easily from equation (8), as for any finite m, $f(n, m) \xrightarrow[n \to \infty]{} 0$, thus the error term $\Phi(-f(n, m))$ approaches $1/2$. on the other hand, when $\boldsymbol{\mu}$ is known the error approaches zero when n increases, since $\|\boldsymbol{\mu}\| \xrightarrow[n \to \infty]{} \infty$ while the variance is fixed. Thus, from corollary 1 we know that by using $m = n$ the error approaches to zero when m grows, and our experiments show that it drops very fast (for $m \sim 20$ it is already below 5% and for $m \sim 300$ below 1%).

Example 4. In this example we show that the property $\mu_j \xrightarrow[j \to \infty]{} 0$ does not guarantee that feature selection can improve classification accuracy. Define $s_n = \sum_{j=1}^{n} \mu_j^2$. Let μ_j be such that $s_n = n^\alpha$ with $\frac{1}{2} < \alpha < 1$. Since $\alpha < 1$, it follows that indeed $\mu_j \to 0$. On the other hand, by substituting in (7) we get:

$$f(n,m) = \frac{n^\alpha}{\sqrt{\frac{n}{m} + n^\alpha}} = \frac{n^{\alpha/2}}{\sqrt{\frac{n^{1-\alpha}}{m} + 1}}$$

thus for $n > m^{\alpha - 1}$,

$$f(n,m) \geq \frac{n^{\alpha/2}}{\sqrt{2\frac{n^{1-\alpha}}{m}}} = n^{\alpha - 1/2} \sqrt{\frac{m}{2}}$$

and since $\alpha > 1/2$, we have that $f(n,m) \xrightarrow[n \to \infty]{} \infty$ and thus using all the features is optimal. We can see that the intuitive relation between the value of the smallest feature we want to take and $\frac{1}{\sqrt{m}}$ that raised in the previous examples does not hold here. this demonstrate that thinking on this value as proportional to $\frac{1}{\sqrt{m}}$ may be misleading.

5   SVM Performance

So far we have seen that the naive maximum likelihood classifier is impeded by using too many weak features, even when all the features are relevant and independent. However it is worth testing whether a modern and sophisticated classifier such as SVM that was designed to work in very high dimensional spaces can overcome the "peaking phenomenon". For this purpose we tested SVM on the above two Gaussian setting in the following way. We generated a training set with 1000 features, and trained linear[2] SVM on this training set 1000 times, each time using a different number of features. Then we calculated the generalization error of each returned classifier analytically. The performance associated with a given number of features n is the generalization error achieved using n features, averaged over 200 repeats. We used the SVM tool-box by Gavin Cawley [14]. The parameter C was tuned manually to be $C = 0.0001$, the value which is favorable to SVM when all (1000) features are used. The results for the examples described in section 4 for three different choices of training set size are presented in figure 3.

We can see that in this setting SVM suffers from using too many features just like the maximum likelihood classifier[3]. On the other hand, it is clear that in other situations SVM does handle huge number of features well, otherwise it could not be used together with kernels. Therefore, in order to understand why SVM fails here, we need to determine in what way our high dimensional scenario is different from the one caused by using kernels. The assumption which

---

[2] The best classifier here is linear, thus linear kernel is expected to give best results.
[3] This strengthen the observation in [15] (page 384) that additional noise features hurts SVM performance.

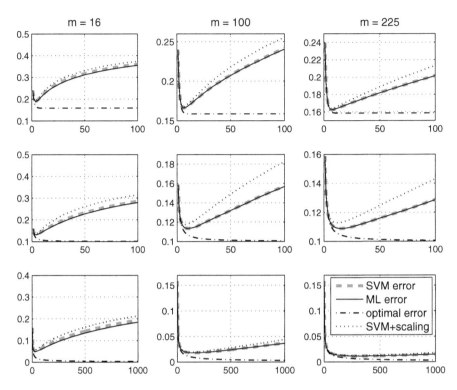

Fig. 3. The error of SVM as a function of the number of features used. The top, middle and bottom rows correspond to μ_j equals $1/\sqrt{2^j}$, $1/j$ and $1/\sqrt{j}$ respectively. The columns correspond to training set sizes of 16, 100 and 225. The SVM error was estimated by averaging over 200 repeats. The graphs show that SVM produces the same qualitative behavior as the ML classifier we used in the analysis and that pre-scaling of the features strengthen the effect of too many features on SVM.

underlies the usage of the large margin principle, namely that the density around the true separator is low, is violated in our first example (example 1), but not for the other two examples (see figure 2). Moreover, if we multiply the μ of example 1 by 2, then the assumption holds and the qualitative behavior of SVM does not change. Hence this cannot be a major factor.

One might suggest that a simple pre-scaling of the features, such as dividing each features by an approximation of its standard deviation[4], might help SVM. Such normalization is useful in many problems, especially when different features are measured in different scales. However, in our specific setting it is not likely that such normalization can improve the accuracy, as it just suppress the more useful features. Indeed, our experiments shows (see figure 3, dotted lines) that the pre-scaling strengthen the effect of too many features on SVM.

---

[4] Given the class, the standard deviation is the same for all the features (equals to 1), but the overall standard deviation is larger for the first features, as in the first coordinates the means of the two classes a more well apart.

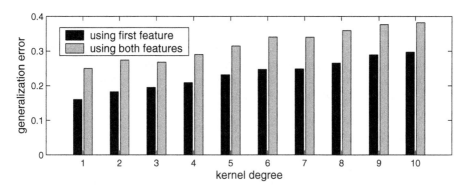

Fig. 4. The effect of the degree of a polynomial kernel on SVM error $\mu = (1, 0)$ and the training set size is $m = 20$. The results were averaged over 200 repeats. The effect of the kernel degree on the difference in accuracy using one or both features is not significant.

One significant difference is that in our setting the features are statistically independent, whereas when the dimension is high due to the usage of kernels, the features are highly correlated. In other words, the use of kernels is equivalent to deterministic mapping of low dimensional space to a high dimensional space. Thus there are many features, but the actual dimension of the embedded manifold is low whereas in our setting the dimension is indeed high. We ran one initial experiment which supported the assumption that this difference was significant in causing the SVM behavior. We used SVM with a polynomial kernel of increasing degrees in the above setting with $\mu = (1, 0)$, i.e. one relevant feature and one irrelevant feature. The results are shown in figure 4. The accuracy declines as the degree increases as expected (since the best model here is linear). However, the effect of the kernel degree on the difference in accuracy using one or both features is not significant, despite the fact that the number of irrelevant features grows exponentially with the kernel degree[5].

6   Discussion

In this work we discussed the relationship between the optimal number of features and the training set size. Using a simple setting of two spherical Gaussians we showed that in some situations SVM does not handle large number of weakly relevant features correctly and achieves suboptimal accuracy, much like a naive classifier. We suggest that the ability of SVM to work well in the presence of huge number of features may be restricted to cases where the underlying distribution is concentrated around a low dimensional manifold, which is the case when kernels are used. However, this issue should be further investigated.

---

[5] When a polynomial kernel of degree k is used, only k features are relevant whereas $2^k - k$ are irrelevant.

In this work we focused on feature selection, but the same fundamental question is relevant for dimensionality reduction as well. In this setting one looks for any conversion of the data to low dimensional subspace that preserves the relevant properties. Thus one should ask in what way the optimal dimension depends on the number of training instances. We do not have definitive formulation for this question yet, but we expect that a similar trade-off can be found.

References

1. J. R. Quinlan. Induction of decision trees. In Jude W. Shavlik and Thomas G. Dietterich, editors, *Readings in Machine Learning*. Morgan Kaufmann, 1990. Originally published in *Machine Learning* 1:81–106, 1986.
2. K. Kira and L. Rendell. A practical approach to feature selection. In *Proc. 9th International Workshop on Machine Learning*, pages 249–256, 1992.
3. H. Almuallim and T. G. Dietterich. Learning with many irrelevant features. In *Proceedings of the Ninth National Conference on Artificial Inte lligence (AAAI-91)*, volume 2, pages 547–552, Anaheim, California, 1991. AAAI Press.
4. Daphne Koller and Mehran Sahami. Toward optimal feature selection. In *International Conference on Machine Learning*, pages 284–292, 1996.
5. R. Gilad-Bachrach, A. Navot, and N. Tishby. Margin based feature selection - theory and algorithms. In *Proc. 21st International Conference on Machine Learning (ICML)*, pages 337–344, 2004.
6. I. Guyon and A. Elisseeff. An introduction to variable and feature selection. *Journal of Machine Learnig Research*, pages 1157–1182, Mar 2003.
7. A. K. Jain and W. G. Waller. On the optimal number of features in the classification of multivariate gaussian data. *Pattern Recognition*, 10:365–374, 1978.
8. G. V. Trunk. A problem of dimensionality: a simple example. *IEEE Transactions on pattern analysis and machine intelligence*, PAMI-1(3):306–307, July 1979.
9. S. Raudys and V. Pikelis. On dimensionality, sample size, classification error and complexity of classification algorithm in pattern recognition. *IEEE Transactions on pattern analysis and machine intelligence*, PAMI-2(3):242–252, May 1980.
10. J. Hua, Z. Xiong, and E. R. Dougherty. Determination of the optimal number of features for quadratic discriminant analysis via normal approximation to the discriminant distribution. *Pattern Recognition*, 38(3):403–421, March 2005.
11. A. K. Jain and W. G. Waller. On the monotonicity of the performance of bayesian classifiers. *IEEE transactions on Information Theory*, 24(3):392–394, May 1978.
12. B. Boser, I. Guyon, and V. Vapnik. Optimal margin classifiers. In *In Fifth Annual Workshop on Computational Learning Theory*, pages 144–152, 1992.
13. T. W. Anderson. Classification by multivariate analysis. *Psychometria*, 16:31–50, 1951.
14. G. C. Cawley. MATLAB support vector machine toolbox (v0.55β) [`http://theoval.sys.uea.ac.uk/~gcc/svm/toolbox`]. University of East Anglia, School of Information Systems, Norwich, Norfolk, U.K. NR4 7TJ, 2000.
15. T. Hastie, R. Tibshirani, and J. Friedman. *The Elements of Statistical Learning*. Springer, 2001.

Class-Specific Subspace Discriminant Analysis for High-Dimensional Data

Charles Bouveyron[1,2], Stéphane Girard[1], and Cordelia Schmid[2]

[1] LMC – IMAG, BP 53, Université Grenoble 1,
38041 Grenoble, Cedex 9 – France
charles.bouveyron@imag.fr, stephane.girard@imag.fr
[2] LEAR – INRIA Rhône-Alpes, 655 avenue de l'Europe, Montbonnot,
38334 Saint-Ismier, Cedex – France
cordelia.schmid@inrialpes.fr

Abstract. We propose a new method for discriminant analysis, called High Dimensional Discriminant Analysis (HDDA). Our approach is based on the assumption that high dimensional data live in different subspaces with low dimensionality. We therefore propose a new parameterization of the Gaussian model to classify high-dimensional data. This parameterization takes into account the specific subspace and the intrinsic dimension of each class to limit the number of parameters to estimate. HDDA is applied to recognize object parts in real images and its performance is compared to classical methods.

Keywords: Discriminant analysis, class-specific subspaces, dimension reduction, regularization.

1 Introduction

Many scientific domains need to analyze data which are increasingly complex. For example, visual descriptors used in object recognition are often high-dimensional and this penalizes classification methods and consequently recognition. In high-dimensional feature spaces, the performance of learning methods suffers from the *curse of dimensionality* [1] which deteriorates both classification accuracy and efficiency. Popular classification methods are based on a Gaussian model and show a disappointing behavior when the size of the training dataset is too small compared to the number of parameters to estimate. To avoid overfitting, it is therefore necessary to find a balance between the number of parameters to estimate and the generality of the model. In this paper we propose a Gaussian model which determines the specific subspace in which each class is located and therefore limits the number of parameters to estimate. The maximum likelihood method is used for parameter estimation and the intrinsic dimension of each class is determined automatically with the scree-test of Cattell. This allows to derive a robust discriminant analysis method in high-dimensional spaces, called High Dimensional Discriminant Analysis (HDDA). It is possible to make additional assumptions on the model to further limit the number of parameters. We can

C. Saunders et al. (Eds.): SLSFS 2005, LNCS 3940, pp. 139–150, 2006.
© Springer-Verlag Berlin Heidelberg 2006

assume that classes are spherical in their subspaces and it is possible to fix some parameters to be common between classes. A comparison with standard discriminant analysis methods on a recently proposed dataset [4] shows that HDDA outperforms them.

This paper is organized as follows. Section 2 presents the discrimination problem and existing methods to regularize discriminant analysis in high-dimensional spaces. Section 3 introduces the theoretical framework of HDDA and, in section 4, some particular cases are studied. Section 5 is devoted to the inference aspects. Our method is then compared to classical methods on a real image dataset in section 6.

2   Discriminant Analysis Framework

2.1   Discrimination Problem

The goal of discriminant analysis is to assign an observation $x \in \mathbb{R}^p$ with unknown class membership to one of k classes $C_1, ..., C_k$ known *a priori*. For this, we have a learning dataset $A = \{(x_1, c_1), ..., (x_n, c_n)/x_j \in \mathbb{R}^p$ and $c_j \in \{1, ..., k\}\}$, where the vector x_j contains p explanatory variables and c_j indicates the index of the class of x_j. The optimal decision rule, called *Bayes decision rule*, assigns the observation x to the class C_{i*} which has the *maximum a posteriori* probability. This is equivalent to minimize a cost function $K_i(x)$, *i.e.*, $i^* = \mathrm{argmin}_{i=1,...,k}\{K_i(x)\}$, with $K_i(x) = -2\log(\pi_i f_i(x))$, where π_i is the *a priori* probability of class C_i and $f_i(x)$ denotes the class conditional density of x, $\forall i = 1, ..., k$. For instance, assuming that $f_i(x)$ is a Gaussian density leads to the well known Linear Discriminant Analysis (LDA) and Quadratic Discriminant Analysis (QDA) methods.

2.2   Dimension Reduction and Parsimonious Models

In high dimensional spaces, the majority of classification methods shows a disappointing behavior when the size of the training dataset is too small compared to the number of parameters to estimate. To avoid overfitting, it is therefore necessary to reduce the number of parameters. This is possible by either reducing the dimension of the data or by using a parsimonious model with additional assumptions on the model.

Dimension reduction. Many methods use global dimension reduction techniques to overcome problems due to high-dimensionality. A widely used solution is to reduce the dimensionality of the data before using a classical discriminant analysis method. The dimension reduction can be done using Principal Components Analysis (PCA) or a feature selection technique. It is also possible to reduce the data dimension with classification as a goal by using Fisher Discriminant Analysis (FDA) which projects the data on the $(k-1)$ discriminant axes and then classifies the projected data. The dimension reduction is often advantageous in terms of performance but loses information which could be discriminant due to the fact that most approaches are global and not designed for classification.

Parsimonious models. Another solution is to use a model which requires the estimation of fewer parameters. The parsimonious models used most often involve an identical covariance matrix for all classes (used in LDA), *i.e.*, $\forall i$, $\Sigma_i = \Sigma$, or a diagonal covariance matrix, *i.e.*, $\Sigma_i = \text{diag}(\sigma_{i1}, ..., \sigma_{ip})$. Other approaches propose new parameterizations of the Gaussian model in order to find different parsimonious models. For example, Regularized Discriminant Analysis [6] uses two regularization parameters to design an intermediate classifier between QDA and LDA. The Eigenvalue Decomposition Discriminant Analysis [2] proposes to re-parameterize the covariance matrices of the classes in their eigenspace. These methods do not allow to efficiently solve the problem of the high-dimensionality, as they do not determine the specific subspaces in which the data of each class live.

3   High Dimensional Discriminant Analysis

The *empty space* phenomenon [9] allows us to assume that high-dimensional data live in different low-dimensional subspaces hidden in the original space. We therefore propose in this section a new parameterization of the Gaussian model which combines a local subspace approach and a parsimonious model.

3.1   The Gaussian Mixture Model

Similarly to classical discriminant analysis, we assume that class conditional densities are Gaussian $\mathcal{N}(\mu_i, \Sigma_i)$, $\forall i = 1, ..., k$. Let Q_i be the orthogonal matrix of eigenvectors of the covariance matrix Σ_i and $\mathcal{B}_i$ be the eigenspace of Σ_i, *i.e.*, the basis of eigenvectors of Σ_i. The class conditional covariance matrix Δ_i is then defined in the basis $\mathcal{B}_i$ by $\Delta_i = Q_i^t \Sigma_i Q_i$. Thus, Δ_i is diagonal and made of eigenvalues of Σ_i. We assume in addition that Δ_i has only two different eigenvalues $a_i > b_i$:

$$\Delta_i = \begin{pmatrix} \begin{array}{|ccc|} \hline a_i & & 0 \\ & \ddots & \\ 0 & & a_i \\ \hline \end{array} & \mathbf{0} \\ \mathbf{0} & \begin{array}{|ccc|} \hline b_i & & 0 \\ & \ddots & \\ 0 & & b_i \\ \hline \end{array} \end{pmatrix} \begin{array}{l} \left.\vphantom{\begin{array}{c}a\\b\\c\end{array}}\right\} \ d_i \\ \\ \left.\vphantom{\begin{array}{c}a\\b\\c\end{array}}\right\} \ (p - d_i) \end{array}$$

Let $\mathbb{E}_i$ be the affine space generated by the eigenvectors associated with the eigenvalue a_i with $\mu_i \in \mathbb{E}_i$, and let $\mathbb{E}_i^\perp$ be $\mathbb{E}_i \oplus \mathbb{E}_i^\perp = \mathbb{R}^p$ with $\mu_i \in \mathbb{E}_i^\perp$. Thus, the class C_i is both spherical in $\mathbb{E}_i$ and in $\mathbb{E}_i^\perp$. Let $P_i(x) = \tilde{Q}_i \tilde{Q}_i^t (x - \mu_i) + \mu_i$ be the projection of x on $\mathbb{E}_i$, where $\tilde{Q}_i$ is made of the d_i first columns of Q_i and supplemented by zeros. Similarly, let $P_i^\perp(x) = (Q_i - \tilde{Q}_i)(Q_i - \tilde{Q}_i)^t (x - \mu_i) + \mu_i$ be the projection of x on $\mathbb{E}_i^\perp$. Figure 1 summarizes these notations.

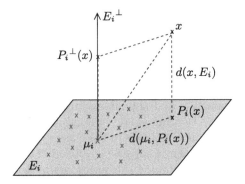

Fig. 1. The subspaces $\mathbb{E}_i$ and $\mathbb{E}_i^\perp$ of the class C_i

3.2   Decision Rule and *a Posteriori* Probability

Deriving the Bayes decision rule with the model described in the previous section yields the decision rule of High Dimensional Discriminant Analysis (HDDA).

Theorem 1. *Bayes decision rule yields the decision rule δ^+ which classifies x as the class C_{i^*} such that $i^* = \mathrm{argmin}_{i=1,\ldots,k}\{K_i(x)\}$ where K_i is defined by:*

$$K_i(x) = \frac{1}{a_i}\|\mu_i - P_i(x)\|^2 + \frac{1}{b_i}\|x - P_i(x)\|^2 + d_i\log(a_i) + (p - d_i)\log(b_i) - 2\log(\pi_i).$$

Proof. We derive Bayes decision rule for the Gaussian model presented in section 3.1. Writing f_i with the class conditional covariance matrix Δ_i gives:

$$-2\log(f_i(x)) = (x - \mu_i)^t(Q_i\Delta_iQ_i^t)^{-1}(x - \mu_i) + \log(\det \Delta_i) + p\log(2\pi).$$

Moreover, $Q_i^tQ_i = Id$ and consequently:

$$-2\log(f_i(x)) = \left[Q_i^t(x - \mu_i)\right]^t \Delta_i^{-1}\left[Q_i^t(x - \mu_i)\right] + \log(\det \Delta_i) + p\log(2\pi).$$

Given the structure of Δ_i, we obtain:

$$-2\log(f_i(x)) = \frac{1}{a_i}\|\tilde{Q}_i^{\,t}(x - \mu_i)\|^2 + \frac{1}{b_i}\|(Q_i - \tilde{Q}_i)^t(x - \mu_i)\|^2$$
$$+ \log(\det \Delta_i) + p\log(2\pi).$$

Using definitions of P_i and $P_i^\perp$ and in view of figure 1, we obtain:

$$-2\log(f_i(x)) = \frac{1}{a_i}\|\mu_i - P_i(x)\|^2 + \frac{1}{b_i}\|x - P_i(x)\|^2 + \log(\det \Delta_i) + p\log(2\pi).$$

The relation $\log(\det \Delta_i) = d_i\log(a_i) + (p - d_i)\log(b_i)$ concludes the proof. $\square$

The *a posteriori* probability $P(x \in C_i|x)$ measures the probability that x belongs to C_i and allows to identify dubiously classified points. Basing on Bayes' formula, we can write: $P(x \in C_i|x) = 1/\sum_{l=1}^k \exp\left(\frac{1}{2}(K_i(x) - K_l(x))\right).$

4   Particular Rules

By allowing some of the HDDA parameters to be common between classes, we obtain particular rules which correspond to different types of regularization, some of which are easily geometrically interpretable. Due to space restrictions, we present only the two most important particular cases: HDDAi and HDDAh. In order to interpret these particular decision rules, the following notations are useful: $\forall i = 1, ..., k$, $a_i = \frac{\sigma_i^2}{\alpha_i}$ and $b_i = \frac{\sigma_i^2}{(1-\alpha_i)}$ with $\alpha_i \in]0, 1[$ and $\sigma_i > 0$.

4.1   Isometric Decision Rule (HDDAi)

Here, the following additional assumptions are made: $\forall i = 1, ..., k$, $\alpha_i = \alpha$, $\sigma_i = \sigma$, $d_i = d$ and $\pi_i = \pi_*$. In this case, the classes are isometric.

Proposition 1. *Under these assumptions, the decision rule classifies x as the class C_{i*} such that $i^* = \mathrm{argmin}_{i=1,...,k}\{K_i(x)\}$ where K_i is defined by:*

$$K_i(x) = \frac{1}{\sigma^2}\left(\alpha\|\mu_i - P_i(x)\|^2 + (1-\alpha)\|x - P_i(x)\|^2\right).$$

For particular values of α, HDDAi has simple geometrical interpretations:

- Case $\alpha = 0$: HDDAi assigns x to the class C_{i*} if $\forall i = 1, ..., k$, $d(x, \mathbb{E}_{i*}) < d(x, \mathbb{E}_i)$. From a geometrical point of view, the decision rule assigns x to the class associated with the closest subspace $\mathbb{E}_i$.
- Case $\alpha = 1$: HDDAi assigns x to the class C_{i*} if $\forall i = 1, ..., k$, $d(\mu_{i*}, P_{i*}(x)) < d(\mu_i, P_i(x))$. It means that the decision rule assigns x to the class for which the mean is closest to the projection of x on the subspace.
- Case $0 < \alpha < 1$: the decision rule assigns x to the class realizing a compromise between the two previous cases. The estimation of α is discussed in the following section.

4.2   Homothetic Decision Rule (HDDAh)

This method differs from the previous one by removing the constraint $\sigma_i = \sigma$, and classes are thus homothetic.

Proposition 2. *In this case, the decision rule classifies x as the class C_{i*} such that $i^* = \mathrm{argmin}_{i=1,...,k}\{K_i(x)\}$ where K_i is defined by:*

$$K_i(x) = \frac{1}{\sigma_i^2}(\alpha\|\mu_i - P_i(x)\|^2 + (1-\alpha)\|x - P_i(x)\|^2) + 2p\log(\sigma_i).$$

HDDAh generally favours classes with a large variance. We can observe on Figure 2 that HDDAh favours the blue class which has the largest variance whereas HDDAi gives the same importance to both classes. It assigns to the blue class a point which is far from the axis of the red class, *i.e.*, which does not live in the specific subspace of the red class.

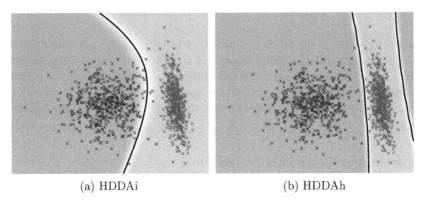

(a) HDDAi   (b) HDDAh

Fig. 2. Decision rules obtained with HDDAi and HDDAh on simulated data

4.3 Removing Constraints on d_i and π_i

The two previous methods assume that d_i and π_i are fixed. However, these assumptions can be too restrictive. If these constraints are removed, it is necessary to add the corresponding terms in $K_i(x)$: if d_i are free, then $d_i \log(\frac{1-\alpha}{\alpha})$ must be added and if π_i are free, then $-2\log(\pi_i)$ must be added.

5 Estimators

The estimators are obtained by Maximum Likelihood (ML) estimation based on the learning dataset. In the following, parameters π_i, μ_i and Σ_i of the class C_i are estimated by their empirical counterparts:

$$\hat{\pi}_i = \frac{n_i}{n}, \ \hat{\mu}_i = \frac{1}{n_i} \sum_{x_j \in C_i} x_j, \ \hat{\Sigma}_i = \frac{1}{n_i} \sum_{x_j \in C_i} (x_j - \hat{\mu}_i)^t (x_j - \hat{\mu}_i),$$

where n_i is the cardinality of the class C_i.

5.1 HDDA Estimators

Assuming for the moment that the d_i are known, we obtain the following ML estimates.

Proposition 3. *The ML estimators of matrices $\tilde{Q}_i$ and parameters a_i and b_i exist and are unique, $\forall i = 1, ..., k$:*

(i) The d_i first columns of $\tilde{Q}_i$ are estimated by the eigenvectors associated to the d_i largest eigenvalues of $\hat{\Sigma}_i$.
(ii) $\hat{a}_i$ is the mean of the d_i largest eigenvalues of $\hat{\Sigma}_i$:

$$\hat{a}_i = \frac{1}{d_i} \sum_{l=1}^{d_i} \lambda_{il},$$

where λ_{il} is the lth largest eigenvalue of $\hat{\Sigma}_i$, *(iii)* $\hat{b}_i$ *is the mean of the* $(p - d_i)$ *smallest eigenvalues of* $\hat{\Sigma}_i$ *and can be written:*

$$\hat{b}_i = \frac{1}{(p - d_i)} \left(\text{Tr}(\hat{\Sigma}_i) - \sum_{l=1}^{d_i} \lambda_{il} \right).$$

Proof. Equation (2.5) of [5] provides the following log-likelihood expression:

$$-2 \log(L_i(x_j \in C_i, \mu_i, \Sigma_i)) = n_i \sum_{l=1}^{p} \left(\log \delta_{il} + \frac{1}{\delta_{il}} q_{il}^t \hat{\Sigma}_i q_{il} \right) + C^{te},$$

with $\delta_{il} = a_i$ if $l \leq d_i$ and $\delta_{il} = b_i$ otherwise. This quantity is to be minimized under the constraint $q_{il}^t q_{il} = 1$, which is equivalent to find a saddle point of the Lagrange function:

$$\mathcal{L}_i = n_i \sum_{l=1}^{p} \left(\log \delta_{il} + \frac{1}{\delta_{il}} q_{il}^t \hat{\Sigma}_i q_{il} \right) - \sum_{l=1}^{p} \theta_{il} (q_{il}^t q_{il} - 1),$$

where θ_{il} are the Lagrange multipliers. The derivative with respect to a_i is:

$$\frac{\partial \mathcal{L}_i}{\partial a_i} = \frac{n_i d_i}{a_i} - \frac{n_i}{a_i^2} \sum_{l=1}^{d_i} q_{il}^t \hat{\Sigma}_i q_{il},$$

and the condition $\frac{\partial \mathcal{L}_i}{\partial a_i} = 0$ implies that:

$$\hat{a}_i = \frac{1}{d_i} \sum_{l=1}^{d_i} q_{il}^t \hat{\Sigma}_i q_{il}. \tag{1}$$

In the same manner, the partial derivative of $\mathcal{L}_i$ with respect to b_i is:

$$\frac{\partial \mathcal{L}_i}{\partial b_i} = \frac{n_i (p - d_i)}{b_i} - \frac{n_i}{b_i^2} \sum_{l=d_i+1}^{p} q_{il}^t \hat{\Sigma}_i q_{il},$$

and the condition $\frac{\partial \mathcal{L}_i}{\partial b_i} = 0$ implies that:

$$\hat{b}_i = \frac{1}{(p - d_i)} \sum_{l=d_i+1}^{p} q_{il}^t \hat{\Sigma}_i q_{il} = \frac{1}{(p - d_i)} \left(\text{Tr}(\hat{\Sigma}_i) - \sum_{l=1}^{d_i} q_{il}^t \hat{\Sigma}_i q_{il} \right). \tag{2}$$

In addition, the gradient of $\mathcal{L}_i$ with respect to q_{il} is $\forall l \leq d_i$:

$$\nabla_{q_{il}} \mathcal{L}_i = 2 \frac{n_i}{\delta_{il}} \hat{\Sigma}_i q_{il} - 2 \theta_{il} q_{il},$$

and by multiplying this quantity on the left by q_{il}^t, we obtain:

$$q_{il}^t \nabla_{q_{il}} \mathcal{L}_i = 0 \Leftrightarrow \theta_{il} = \frac{n_i}{\delta_{il}} q_{il}^t \hat{\Sigma}_i q_{il}.$$

Consequently, $\hat{\Sigma}_i q_{il} = \frac{\theta_{il}\delta_{il}}{n_i} q_{il}$ which means that q_{il} is the eigenvector of $\hat{\Sigma}_i$ associated to the eigenvalue $\lambda_{il} = \frac{\theta_{il}\delta_{il}}{n_i}$. Replacing in (1) and (2), we obtain the ML estimators for a_i and b_i. Vectors q_{il} being eigenvectors of $\hat{\Sigma}_i$ which is a symmetric matrix, this implies that $q_{il}^t q_{ih} = 0$ if $h \neq l$. In order to minimize the quantity $-2 \log L_i$ at the *optimum*, $\hat{a}_i$ must be as large as possible. Thus, the d_i first columns of Q_i must be the eigenvectors associated to the d_i largest eigenvalues of $\hat{\Sigma}_i$. □

Note that the decision rule of HDDA requires only the estimation of the matrix $\tilde{Q}_i$ instead of the entire Q_i and this reduces significantly the number of parameters to estimate. For example, if we consider 100-dimensional data, 4 classes and common intrinsic dimensions d_i equal to 10, HDDA estimates only 4 323 parameters whereas QDA estimates 20 603 parameters.

5.2   HDDAi Estimators

Proposition 4. *The ML estimators of parameters α and σ exist and are unique:*

$$\hat{\alpha} = \frac{\hat{b}}{\hat{a} + \hat{b}}, \quad \hat{\sigma}^2 = \frac{\hat{a}\hat{b}}{\hat{a} + \hat{b}},$$

with $\hat{a} = \frac{\sum_{i=1}^{k} n_i \sum_{l=1}^{d_i} \lambda_{il}}{np\gamma}, \hat{b} = \frac{\sum_{i=1}^{k} n_i \left(\mathrm{Tr}(\hat{\Sigma}_i) - \sum_{l=1}^{d_i} \lambda_{il} \right)}{np(1-\gamma)}$ *where* $\gamma = \frac{1}{np} \sum_{i=1}^{k} n_i d_i$ *and* λ_{il} *is the lth largest eigenvalue of* $\hat{\Sigma}_i$.

Proof. In this case, the log-likelihood expression is:

$$-2 \log(L) = \sum_{i=1}^{k} n_i \sum_{l=1}^{p} \left(\log \delta_{il} + \frac{1}{\delta_{il}} \lambda_{il} \right) + C^{te},$$

where $\delta_{il} = a$ if $l \leq d_i$ and b otherwise. One can write:

$$-2 \frac{\partial}{\partial a} \log(L) = 0 \Leftrightarrow \sum_{i=1}^{k} n_i \sum_{l=1}^{d_i} \left(\frac{1}{a} - \frac{1}{a^2} \lambda_{il} \right) = 0 \Leftrightarrow \hat{a} = \frac{\sum_{i=1}^{k} n_i \sum_{l=1}^{d_i} \lambda_{il}}{np\gamma},$$

with $\gamma = \frac{1}{np} \sum_{i=1}^{k} n_i d_i$. Similarly,

$$-2 \frac{\partial}{\partial b} \log(L) = 0 \Leftrightarrow \hat{b} = \frac{\sum_{i=1}^{k} n_i \left(\mathrm{Tr}(\hat{\Sigma}_i) - \sum_{l=1}^{d_i} \lambda_{il} \right)}{np(1-\gamma)}.$$

Replacing these estimates in expressions of α and σ concludes the proof. □

5.3   HDDAh Estimators

Proposition 5. *The ML estimate of α has the following formulation according to σ_i, $\forall i = 1, ..., k$:*

$$\hat{\alpha}(\sigma_1, ..., \sigma_k) = \frac{(\Lambda + 1) - \sqrt{\Delta}}{2\Lambda},$$

with the notations:

$$\Delta = (\Lambda + 1)^2 - 4\Lambda\gamma, \; \Lambda = \frac{1}{np} \sum_{i=1}^{k} \frac{n_i}{\sigma_i^2} \left(2 \sum_{l=1}^{d_i} \lambda_{il} - \mathrm{Tr}(\hat{\Sigma}_i) \right),$$

and the ML estimate of σ_i^2 has the following formulation according to α:

$$\forall i = 1, ..., k, \; \hat{\sigma_i^2}(\alpha) = \frac{1}{p} \left((2\alpha - 1) \sum_{l=1}^{d_i} \lambda_{il} + (1 - \alpha) \mathrm{Tr}(\hat{\Sigma}_i) \right).$$

Proof. In this case, one can write:

$$-2\log(L) = \sum_{i=1}^{k} n_i \left[2p \log \sigma_i - d_i \log \alpha - (p - d_i) \log(1 - \alpha) \right.$$

$$\left. + \frac{1}{\sigma_i^2} \left((2\alpha - 1) \sum_{l=1}^{d_i} \lambda_{il} + (1 - \alpha) \mathrm{Tr}(\hat{\Sigma}_i) \right) \right].$$

Therefore,

$$\frac{\partial}{\partial \alpha} \log(L) = 0 \Leftrightarrow \sum_{i=1}^{k} n_i \left(-\frac{d_i}{\alpha} + \frac{(p - d_i)}{(1 - \alpha)} + \frac{2 \sum_{l=1}^{d_i} \lambda_{il}}{\sigma_i^2} - \frac{\mathrm{Tr}(\hat{\Sigma}_i)}{\sigma_i^2} \right) = 0,$$

$$\Leftrightarrow np \left(-\frac{\gamma}{\alpha} + \frac{(1 - \gamma)}{(1 - \alpha)} + \Lambda \right) = 0,$$

where $\gamma = \frac{1}{np} \sum_{i=1}^{k} n_i d_i$ and $\Lambda = \frac{1}{np} \sum_{i=1}^{k} \frac{n_i}{\sigma_i^2} \left(2 \sum_{l=1}^{d_i} \lambda_{il} - \mathrm{Tr}(\hat{\Sigma}_i) \right)$. Thus,

$$\frac{\partial}{\partial \alpha} \log(L) = 0 \Leftrightarrow \psi(\alpha) = \Lambda \alpha^2 - (\Lambda + 1)\alpha + \gamma = 0.$$

The discriminant of the previous equation is $\Delta = (\Lambda + 1 - 2\gamma)^2 + 4\gamma(1 - \gamma)$ with $\gamma < 1$ and consequently $\Delta > 0$. By remarking that $\psi(0) = \gamma > 0$ and $\psi(1) = \gamma - 1 < 0$, one can conclude that the solution is in $[0, 1]$ and is the smallest of both solutions of $\frac{\partial}{\partial \alpha} \log(L) = 0$. In addition,

$$\frac{\partial}{\partial \sigma_i} \log(L) = 0 \Leftrightarrow \sigma_i^2 = \frac{1}{p} \left((2\alpha - 1) \sum_{l=1}^{d_i} \lambda_{il} + (1 - \alpha) \mathrm{Tr}(\hat{\Sigma}_i) \right),$$

and thus provides the expression of σ_i^2 according to α. $\qquad\square$

Note that the estimators of both α and σ_i are not explicit and thus they should be computed using an iterative procedure.

5.4   Estimation of the Intrinsic Dimension

The estimation of the dataset intrinsic dimension is a difficult problem which does not have an explicit solution. If the dimensions d_i are common between classes, *i.e.*, $\forall i = 1, ..., k$, $d_i = d$, we determine by cross-validation the dimension d which maximizes the correct classification rate on the learning dataset. Otherwise, we use an approach based on the eigenvalues of the class conditional covariance matrix $\hat{\Sigma}_i$. The jth eigenvalue of $\hat{\Sigma}_i$ corresponds to the fraction of the full variance carried by the jth eigenvector of $\hat{\Sigma}_i$. We therefore estimate the class specific dimension d_i, $i = 1, ..., k$, with the empirical method scree-test of Cattell [3] which analyzes the differences between eigenvalues in order to find a break in the scree. The selected dimension is the one for which the subsequent differences are smaller than a threshold t. In our experiments, the threshold t is chosen by cross-validation. We also used the probabilistic criterion BIC [8] which gave very similar dimension choices. In our experiments, the estimated intrinsic dimensions are on average 10 whereas the dimension of the original space is 128.

6   Experimental Results

Object recognition is one of the most challenging problems in computer vision. Many successful object recognition approaches use local images descriptors. However, local descriptors are high-dimensional and this penalizes classification methods and consequently recognition. HDDA seems therefore well adapted to this problem. In this section, we use HDDA to recognize object parts in images.

6.1   Protocol and Data

For our experiments, small scale-invariant regions are detected on each image and they are characterized by the local SIFT descriptor [7]. We extracted SIFT descriptors of dimension 128 for 100 motorbike images from a recently proposed visual recognition database [4]. For these local descriptors, we selected 2000 descriptors representing 4 classes: wheels, seat, handlebars and background. The learning and the test dataset contain respectively 500 and 1500 descriptors. The pre-processed data are available at `http://lear.inrialpes.fr/~bouveyron/data/data_swc.tgz`. We compared HDDA to the following classical discriminant analysis methods: Linear Discriminant Analysis (LDA), Fisher Discriminant Analysis (FDA) and Support Vector Machines with a RBF kernel (SVM). The parameters t of HDDA and γ of SVM are estimated by cross-validation on the learning dataset.

6.2   Recognition Results

Figure 3 shows recognition results obtained using HDDA methods and state-of-the-art methods with respect to the number of descriptors classified as positive.To obtain the plots we vary the decision boundary between object classes and background, *i.e.*, we change the posterior probabilities provided by generative methods and, for SVM, we vary the parameter C. On the left, only the

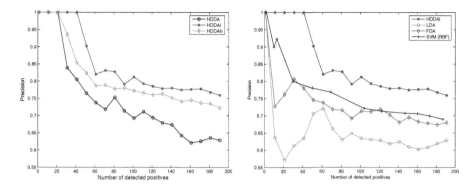

Fig. 3. Classification results for the "motorbike" recognition: comparison between HDDA methods (left) and between HDDAi and state-of-the-art methods (right)

Fig. 4. Recognition of "motorbike" parts using HDDAi (top) and SVM (bottom). The colors blue, red and green are respectively associated to handlebars, wheels and seat.

descriptors with the highest probabilities to belong to the object are used. As a results only a small number of descriptors are classified as positive and their precision (number of correct over total number) is high.

The left plot shows that HDDAi is more efficient than other HDDA methods for this application. This is due to the fact that parameters b_i are common in HDDAi, *i.e.*, the noise is common between classes. More extensive experiments have confirmed that HDDA with common b_i performs in general well for our data. The right plot compares HDDAi to SVM, LDA and FDA. First of all, we observe that the results for LDA (without dimension reduction) are unsatisfying. The results for FDA show that global dimension reduction improves the results. Furthermore, HDDAi obtains better results than SVM and FDA and this demonstrates that our class-specific subspace approach is a good way to classify high-dimensional data. Note that HDDA and HDDAh are also more precise than these three methods when the number of detected positives is small. A comparison with a SVM with a quadratic kernel did not improve the results over the RBF kernel.

Figure 4 presents recognition results for a few test images. These results confirm that HDDAi gives better recognition results than SVM, *i.e.*, the classification errors are significantly lower for HDDAi than for SVM. For example, in the 3rd column of Figure 4, HDDA recognizes the motorbike parts without error whereas SVM makes five errors. In addition, training time for HDDA is as fast as other generative methods and 7 times faster than SVM. Note that recognition time is negligible for all methods.

7   Conclusion

We presented a parameterization of the Gaussian model to classify high-dimensional data in a supervised framework. This model results in a robust and fast discriminant analysis method. We successfully used this method to recognize object parts in natural images. An extension of this work is to use the statistical model of HDDA to adapt the method to unsupervised classification.

Acknowledgment

This work was supported by the French department of research through the *ACI Masse de données* (MoViStaR project).

References

1. R. Bellman. *Dynamic Programing*. Princeton University Press, 1957.
2. H. Bensmail and G. Celeux. Regularized Gaussian Discriminant Analysis Through Eigenvalue Decomposition. *Journal of the American Statistical Association*, 91: 1743–1748, 1996.
3. R. B. Catell. The scree test for the number of factors. *Multivariate Behavioral Research*, 1:140–161, 1966.
4. R. Fergus, P. Perona, and A. Zisserman. Object Class Recognition by Unsupervised Scale-Invariant Learning. In *IEEE Conference on Computer Vision and Pattern Recognition*, pages 264–271, 2003.
5. B. W. Flury. Common Principal Components in k groups. *Journal of the American Statistical Association*, 79:892–897, 1984.
6. J.H. Friedman. Regularized Discriminant Analysis. *Journal of the American Statistical Association*, 84:165–175, 1989.
7. D. Lowe. Distinctive image features from scale-invariant keypoints. *International Journal of Computer Vision*, 60(2):91–110, 2004.
8. G. Schwarz. Estimating the dimension of a model. *The Annals of Statistics*, 6: 461–464, 1978.
9. D. Scott and J. Thompson. Probability density estimation in higher dimensions. In *Proceedings of the Fifteenth Symposium on the Interface, North Holland-Elsevier Science Publishers*, pages 173–179, 1983.

Incorporating Constraints and Prior Knowledge into Factorization Algorithms - An Application to 3D Recovery

Amit Gruber and Yair Weiss

School of Computer Science and Engineering,
The Hebrew University of Jerusalem, Jerusalem, Israel 91904
{amitg, yweiss}@cs.huji.ac.il

Abstract. Matrix factorization is a fundamental building block in many computer vision and machine learning algorithms. In this work we focus on the problem of "structure from motion" in which one wishes to recover the camera motion and the 3D coordinates of certain points given their 2D locations. This problem may be reduced to a low rank factorization problem. When all the 2D locations are known, singular value decomposition yields a least squares factorization of the measurements matrix. In realistic scenarios this assumption does not hold: some of the data is missing, the measurements have correlated noise, and the scene may contain multiple objects. Under these conditions, most existing factorization algorithms fail while human perception is relatively unchanged. In this work we present an EM algorithm for matrix factorization that takes advantage of prior information and imposes strict constraints on the resulting matrix factors. We present results on challenging sequences.

1 Introduction

The problem of "structure from motion" (SFM) has been studied extensively [15, 14, 11, 10, 2] in computer vision: Given the 2D locations of points along an image sequence, the goal is to retrieve the 3D locations of the points. Under simplified camera models, this problem reduces to the problem of matrix factorization [15].

Using SVD, the correct 3D structure can be recovered even if the measurements matrix is contaminated with significant amounts of noise and if the number of frames is small [15].

However, in realistic situations the measurement matrix will have *missing* entries, due to occlusions or due to inaccuracies of the tracking algorithm. A number of algorithms for factorization with missing data [15, 10, 14, 2] have been suggested. While some of these algorithms obtain good results when the data is noiseless, in the presence of even small amounts of noise these algorithms fail.

The problem becomes much harder when the input sequence contains multiple objects with different motions. Not only do we need to recover camera parameters and scene geometry, but we also need to decide which data points should be grouped together. This problem was formulated as a matrix factorization problem by Costeira-Kanade [3]. They suggested to compute an affinity matrix related to the singular value

C. Saunders et al. (Eds.): SLSFS 2005, LNCS 3940, pp. 151–162, 2006.
© Springer-Verlag Berlin Heidelberg 2006

decomposition of the measurements matrix. Then they decide whether two points have the same motion or not by inspecting if some entries of this affinity matrix are zero or not (Gear [5] and Zelnik-Manor et al. [17] follow a similar approach). In the noiseless case these methods perform well, but once even small amounts of noise exist, these methods no longer work since matrix entries that were supposed to be zero are not zero anymore. Furthermore, these methods require some prior knowledge on the rank of the different motions, or linear independence between them.

In this paper we present a framework for matrix factorization capable of incorporating priors and enforcing strict constraints on the desired factorization while handling missing data and correlated noise in the observations. Previous versions of this work were published in [7, 8].

2   Structure from Motion: Problem Formulation and an Algorithm

A set of P feature points in F images are tracked along an image sequence. Let (u_{fp}, v_{fp}) denote image coordinates of feature point p in frame f. Let $W = (w_{ij})$ where $w_{2i-1,j} = u_{ij}$ and $w_{2i,j} = v_{ij}$ for $1 \leq i \leq F$ and $1 \leq j \leq P$.

In the orthographic camera model, points in the 3D world are projected in parallel onto the image plane. For example, if the image coordinate system is aligned with the coordinate system of the 3D world, then a point $P = [X, Y, Z]^T$ is projected to $p = (u, v) = (X, Y)$ (the depth, Z, has no influence on the image). In this model, a camera can undergo rotation, translation, or a combination of the two. W can be written as [15]:

$$[W]_{2F \times P} = [M]_{2F \times 4} [S]_{4 \times P} + [\eta]_{2F \times P} \tag{1}$$

where $M = \begin{bmatrix} M_1 \\ \vdots \\ M_F \end{bmatrix}_{2F \times 4}$ and $S = \begin{bmatrix} X_1 & \cdots & X_P \\ Y_1 & \cdots & Y_P \\ Z_1 & \cdots & Z_P \\ 1 & \cdots & 1 \end{bmatrix}_{4 \times P}$.

Each M_i is a 2×4 matrix that describes camera parameters in the i'th frame. It consists of location and orientation $[M_i]_{2 \times 4} = \begin{bmatrix} m_i^T & d_i \\ n_i^T & e_i \end{bmatrix}$ where m_i and n_i are 3×1 vectors that describe the rotation of the camera; d_i and e_i are scalars describing camera translation[1]. The matrix S contains the 3D coordinates of the feature points, and η is Gaussian noise.

If the elements of the noise matrix η are uncorrelated and of equal variance then we seek a factorization that minimizes the mean squared error between W and MS. This can be solved trivially using the SVD of W. Missing data can be modeled using equation 1 by assuming some elements of the noise matrix η have infinite variance. Obviously SVD is not the solution once we allow different elements of η to have different variances.

---

[1] Note that we do not subtract the mean of each row from it, since in case of missing data the centroids of visible points in different rows of the matrix do not coincide.

2.1   Factorization as Factor Analysis

We seek a factorization of W to M and S that minimizes the weighted squared error $\sum_t [(W_t - M_t S)^T \Psi_t^{-1}(W_t - M_t S)]$, where Ψ_t^{-1} is the inverse covariance matrix of the feature points in frame t.

It is well known that the SVD calculation can be formulated as a limiting case of maximum likelihood (ML) factor analysis [12]. In standard factor analysis we have a set of observations $\{y(t)\}$ that are linear combinations of latent variables $\{x(t)\}$:

$$y(t) = Ax(t) + \eta(t) \qquad (2)$$

with $x(t) \sim N(0, \sigma_x^2 I)$ and $\eta(t) \sim N(0, \Psi_t)$. In the case of a diagonal Ψ_t with constant elements $\Psi_t = \sigma^2 I$ then in the limit $\sigma/\sigma_x \to 0$ the ML estimate for A will give the same answer as the SVD.

Let $A = S^T$. Identifying $y(t)$ with the t'th row of the matrix W and $x(t)$ with the t'th row of M, then equation 1 is equivalent (transposed) to equation 2. Therefore, equation 1 can be solved using the EM algorithm for factor analysis [13] which is a standard algorithm for finding the ML estimate for the matrix A. The EM algorithm consists of two steps: (1) the expectation (or E) step in which expectations are calculated over the latent variables $x(t)$ and (2) the maximization (or M) step in which these expectations are used to maximize the likelihood of the matrix A. The updating equations are:

E step:

$$E(x(t)|y(t)) = \left(\sigma_x^{-2}I + A^T \Psi_t^{-1} A\right)^{-1} A^T \Psi_t^{-1} y(t) \qquad (3)$$

$$V(x(t)|y(t)) = \left(\sigma_x^{-2}I + A^T \Psi_t^{-1} A\right)^{-1} \qquad (4)$$

$$\langle x(t) \rangle = E(x(t)|y(t)) \qquad (5)$$

$$\left\langle x(t)x(t)^T \right\rangle = V(x(t)|y(t)) + \langle x(t) \rangle \langle x(t) \rangle^T \qquad (6)$$

Although in our setting the matrix A must satisfy certain constraints, the E-step (in which the matrix A is assumed to be given from the M-step) remains the same as in standard factor analysis. So far, we assumed no prior on the motion of the camera, i.e. $\sigma_x \to \infty$ and thus $\sigma_x^{-2} \to 0$. In subsection 2.2 we describe how to incorporate priors regarding the motion into the E-step.

M step: In the M step we find the 3D coordinates of a point p denoted by $s_p \in R^3$:

$$s_p = B_p C_p^{-1} \qquad (7)$$

where

$$B_p = \sum_t \left[\Psi_t^{-1}(p, p)(u_{tp} - \langle d_t \rangle) \left\langle m(t)^T \right\rangle \right. \qquad (8)$$

$$\left. + \Psi_t^{-1}(p + P, p + P)(v_{tp} - \langle e_t \rangle) \left\langle n(t) \right\rangle^T \right]$$

$$C_p = \sum_t \left[\Psi_t^{-1}(p, p) \left\langle m(t)m(t)^T \right\rangle \right.$$

$$\left. + \Psi_t^{-1}(p + P, p + P) \left\langle n(t)n(t)^T \right\rangle \right]$$

where the expectations required in the M step are the appropriate subvectors and submatrices of $\langle x(t) \rangle$ and $\langle\, x(t)x(t)^T \,\rangle$.

If we set $\Psi_t^{-1}(p, p) = 0$ when point p is missing in frame t then we obtain an EM algorithm for factorization with missing data. Note that the form of the updates means that we can put any value we wish in the missing elements of y and they will be ignored by the algorithm.

A more realistic noise model for real images is that Ψ_t is *not diagonal* but rather that the noise in the horizontal and vertical coordinates of the same point are correlated with an arbitrary 2×2 inverse covariance matrix. It can be shown that the posterior inverse covariance matrix is $\begin{bmatrix} \sum I_x^2 & \sum I_x I_y \\ \sum I_x I_y & \sum I_y^2 \end{bmatrix}$ (I_x and I_y are the directional derivatives of the image and the sum is taken over a window of fixed size around each pixel). This problem is usually called *factorization with uncertainty* [9, 11]. To consider dependencies between the u and v coordinates of a point, the matrix W can be reshaped (to size $F \times 8$) to have both coordinates in the same row (with a corresponding change in M and S). A non diagonal Ψ_t would express the correlation of the noise in the horizontal and vertical coordinates of the same point. With this representation, it is easy to derive the M step in this case as well. It is similar to equation 7 except that cross terms involving $\Psi_t^{-1}(p, p + P)$ are also involved:

$$s_p = (B_p + B'_p)(C_p + C'_p)^{-1} \tag{9}$$

where

$$B'_p = \sum_t \left[\Psi_t^{-1}(p, p + P)(v_{tp} - \langle e_t \rangle) \langle m(t)^T \rangle \right. \tag{10}$$
$$\left. + \Psi_t^{-1}(p + P, p)(u_{tp} - \langle d_t \rangle) \langle n(t) \rangle^T \right]$$
$$C'_p = \sum_t \left[\Psi_t^{-1}(p, p + P) \langle n(t)m(t)^T \rangle \right.$$
$$\left. + \Psi_t^{-1}(p + P, p) \langle m(t)n(t)^T \rangle \right]$$

Regardless of uncertainty and missing data, the complexity of the EM algorithm grows linearly with the number of feature points and the number of frames.

2.2   Adding Priors on the Desired Factorization

The EM framework allows us to place priors on both structure and motion and to deal with directional uncertainty and missing data. We first show how to place a prior on the motion in the form of temporal coherence. Next we show how to place a prior on the 3D structure of the scene.

Temporal Coherence: The factor analysis algorithm assumes that the latent variables $x(t)$ are independent (figure 1(a)). In SFM this assumption means that the camera locations in different frames are independent and hence permuting the order of the frames makes no difference for the factorization. In almost any video sequence this assumption is wrong. Typically camera location varies smoothly as a function of time (figure 1(b)).

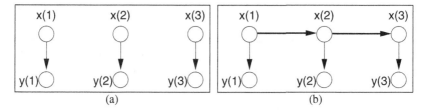

Fig. 1. a. The graphical model assumed by most factorization algorithms for SFM. The camera location $x(t)$ is assumed to be independent of the camera location at any other time step. **b.** The graphical model assumed by our approach. We model temporal coherence by assuming a Markovian structure on the camera location.

Specifically, in this work we use a second order approximation to the motion of the camera (details can be found in [7, 8]). Note that we do not assume that the 2D trajectory of each point is smooth. Rather we assume the 3D trajectory of the camera is smooth.

It is straightforward to derive the EM iterations for a ML estimate of S using the model in figure 1(b). The M step is unchanged from the classical factor analysis. The only change in the E step is that $E(x(t)|y)$ and $V(x(t)|y)$ need to be calculated using a Kalman smoother. We use a standard RTS smoother [6]. Note that the computation of the E step is still linear in the number of frames and datapoints.

Prior on Structure: Up to this point, we have assumed nothing regarding the $3D$ coordinates of the feature points we are trying to reconstruct. The true $3D$ coordinates are considered (a priori) as likely as any other coordinates, even ones that suggest the object is located at an infinite position, or behind the camera, for example. Usually when sequences are acquired for structure reconstruction, the object is located just in front of the camera in the center of the scene, and not at infinity[2]. Therefore, we should prefer reconstructions that place the feature points around certain coordinates in the world, denoted by S_0 (typically X and Y are scattered around zero and Z is finite). We model this preference with the following prior: $\Pr(S) \propto e^{-\lambda \|S - S_0\|_F^2}$, where λ is a parameter that determines the weight of this prior.

Derivation of the modified M-step with the addition of the prior on structure yields (the following modification of equation 9):

$$s_p = (B_p + B'_p)(C_p + C'_p + \lambda(I - S_0))^{-1} \qquad (11)$$

Experimental results show an improvement in reconstruction results in noisy scenes after the addition of this naive prior.

3   Constrained Factorization for Subspace Separation

In dynamic scenes with multiple moving objects, each of the K independent motions has its own motion parameters, M_i^j (a 2×4 matrix describing the jth camera parameters

---

[2] Although for objects to comply with affine model they have to be located relatively far from the camera, they are not placed at infinity.

at time i). Denote by S_j the $4 \times P_j$ matrix of the P_j points moving according to the jth motion component. The matrix formed by taking the locations of points sharing the same motion along the sequence is of rank 4. In other words, the vectors of point locations of points with the same motion form a 4D linear subspace defined by the common motion (in fact this is a $3D$ affine subspace). This is a problem of subspace separation.

Let $\tilde{W}$ be a matrix of observations whose columns are ordered to group together points with the same motion. Then ([3]):

$$\left[\tilde{W}\right]_{2F \times P} = M\tilde{S} = \begin{bmatrix} M_1^1 & \cdots & M_1^K \\ \vdots & & \\ M_F^1 & \cdots & M_F^K \end{bmatrix}_{2F \times 4K} \begin{bmatrix} S_1 & 0 & \cdots & 0 \\ 0 & S_2 & \cdots & 0 \\ \vdots & & & \\ 0 & 0 & \cdots & S_K \end{bmatrix}_{4K \times P} \tag{12}$$

In real sequences, however, measurements are not grouped according to their motion. Therefore, the observation matrix, W, is an arbitrary column permutation of the ordered matrix $\tilde{W}$:

$$W = \tilde{W}\Pi = MS \tag{13}$$

where $S_{4K \times P}$ describes scene structure (with unordered columns) and $\Pi_{P \times P}$ is a column permutation matrix. Hence, the structure matrix S is in general not block diagonal, but rather a column permutation of a block diagonal matrix:

$$S = \tilde{S}\Pi \tag{14}$$

Therefore, in each column of the structure matrix corresponding to a point belonging to the kth motion, only entries $4(k-1) + 1, \ldots, 4k$ can be non-zeros (entry $4k$ always equals one).

Finding a factorization of W to M and S that satisfies this constraint would solve the subspace separation problem: from the indices of the non-zero entries in S we can assign each point to the appropriate motion component.

The constrained factorization problem can be written as a constrained factor analysis problem as follows: By substituting $A = S^T$ and identifying $x(t)$ with the tth row of M, the constrained factorization problem is equivalent to the factor analysis problem of equation 2 where A is subject to the constraints on S^T. We adapt the EM algorithm for single motion presented in the previous section to solve constrained factor analysis problem.

Since the matrix A is assumed to be known in the E step, no change is required in the E step of the algorithm from the previous section. The M step, on the other hand, should be modified to find A that satisfies the constraints.

We modify the M step to find S that is a permuted block diagonal matrix. The columns of S (which are the rows of A) can be found independently on each other (each point is independent on the other points given the motion). We show how to find each of the columns of S that will contain non zeros only in the 4 entries corresponding to its most likely motion. Denote by π_p the motion that maximizes the likelihood for point p and let $\pi = (\pi_1, \ldots, \pi_P)$. Let s_p denote the $3D$ coordinates of point p, and let

$\hat{S}$ denote $[s_1, \ldots, s_P]$ the $3D$ coordinates of all points (S contains both segmentation and geometry information, $\hat{S}$ contains only geometry information).

We look for S that maximizes the expected complete log likelihood (where the expectation is taken over M, the motion parameters of all motion components at all times). Maximizing the expected complete log likelihood is equivalent to minimizing of the expectation of an energy term. In terms of energy minimization, the expectation of the energy due to equation 13 is:

$$E(S) = E(\hat{S}, \pi) = \left\langle E(\hat{S}, \pi, M) \right\rangle_M = \tag{15}$$

$$\sum_p \langle E(s_p, \pi_p, M) \rangle_M =$$

$$\sum_p \sum_t \langle ((W_{t,p} - M_{t,\pi_p} s_p)^T \Psi_{t,p}^{-1} (W_{t,p} - M_{t,\pi_p} s_p)) \rangle_M$$

The energy is the weighted sum of square error of the matrix equation 13. In other words, it is the sum of the error over all the points at all times, weighted by the inverse covariance matrix $\Psi_{t,p}^{-1}$ (the sum over the points is implicit in the vectorial notation of the energy for a single motion at the beginning of section 2.1).

As can be seen from equation 15, $E(S)$ can be represented as a sum of terms $E_p(s_p, \pi_p) = \langle E(s_p, \pi_p, M) \rangle_M$ involving a single point:

$$E(S) = \left\langle E(\hat{S}, \pi, M) \right\rangle_M = \sum_p E_p(s_p, \pi_p) \tag{16}$$

Therefore the minimization of $E(S)$ can be performed by minimizing $E_p(s_p, \pi_p)$ for each point p independent on the others.

Since s_p is unknown, we define

$$E_p(\pi_p) = \min_{s_p} E_p(s_p, \pi_p) \tag{17}$$

And we get

$$\min_{s_p, \pi_p} E_p(s_p, \pi_p) = \min_{\pi_p} \left[\min_{s_p} E_p(s_p, \pi_p) \right] = \min_{\pi_p} E_p(\pi_p) \tag{18}$$

Let $s_p^k = \arg\min_{s_p} E_p(s_p, k)$ for a given k. The value of s_p^k can be computed using one of the equations 7, 9,11, replacing $d_t, m(t), e_t$ and $n(t)$ with $d_t^k, m_k(t), e_t^k$ and $n_k(t)$ respectively. Once all the s_p^k are known, $E_p(k)$ are computed for all k by substituting s_p^k in equation 15. Then we choose $\pi_p = \arg\min_k E_p(k)$. The new value of the pth column of S is all zeros except the four entries $4(\pi_p - 1) + 1, \ldots, 4\pi_p$. Entries $4(\pi_p - 1) + 1, \ldots, 4(\pi_p - 1) + 3$ are set to be s_p^k and entry $4\pi_p$ is set to 1.

By modifying the EM algorithm to deal with constrained factorization we now have an algorithm that is guaranteed to find a factorization where the structure matrix has at most 4 nonzero elements per column, even in the presence of noise (in contrast to [3, 5, 17]). Note that no prior knowledge of the rank of the different motions is needed, neither is any assumption on the linear independence of the different motions.

4   Experiments

In this section we describe the experimental performance of EM for SFM and for motion segmentation. In each case we describe the performance of EM with and without temporal coherence.

4.1   EM for SFM

We evaluate EM for structure from motion compared to ground truth and to previous algorithms for structure from motion with missing data [15, 10, 14, 2]. For [15, 10, 14] we used the Matlab implementation made public by D. Jacobs [3].

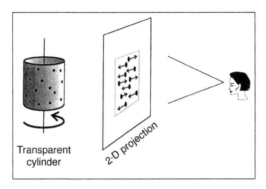

Fig. 2. Synthetic input for evaluation of structure from motion algorithms. A transparent cylinder is rotating around its elongated axis. Points randomly drawn from its surface are projected on the camera plane at each frame. Replotted from [1].

The first input sequence is a synthetic sequence of a transparent rotating cylinder as depicted in figure 2. This sequence (that was first presented in [16]) consists of 100 points uniformly drawn from the cylinder surface. The points are tracked along 20 frames. We checked the performance of the different algorithms in the following cases: (1) full noise free observation matrix , (2) noisy full observation matrix (to create noisy input, the observed image locations were added a Gaussian noise with $\sigma = 0, \ldots, 0.5$), (3) noiseless observations with missing data and (4) noisy observations with missing data.

All algorithms performed well and gave similar results for the full matrix noiseless sequence.

In the fully observed noisy case, factor analysis without temporal coherence gave comparable performance to the algorithm of Tomasi-Kanade, which minimizes $\|MS - W\|_F^2$. When temporal coherence was added, the reconstruction results were improved. The results of Shum's algorithm were similar to Tomasi-Kanade. The algorithms of Jacobs and Brand turned out to be noise sensitive.

In the experiments with missing data, Tomasi-Kanade's algorithm and Shum's algorithm could not handle this pattern of missing data and failed to give any structure. The algorithms of Jacobs and Brand turned out to be noise sensitive.

---

[3] The code is available at http://www.cs.umd.edu/~djacobs

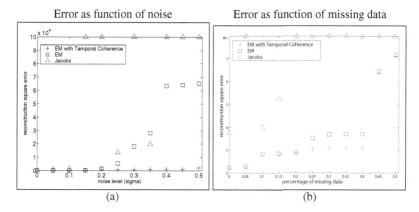

Fig. 3. The graphs depict influence of noise and percentage of missing data on reconstruction results of factor analysis and [10]. The input sequence for these experiments is depicted in figure 2.

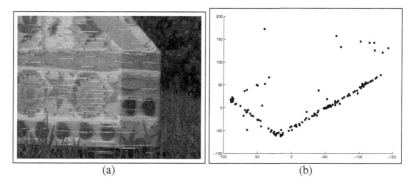

Fig. 4. Results of scene reconstruction from a real sequence: A binder is placed on a rotating surface filmed with a static camera. Our algorithm succeeded in (approximately) obtaining the correct structure while all other algorithms failed. **a.** The first frame of the sequence. **b.** The reconstructed object shown in top view. The 3 lines visible are the outlines of the object. Each of these lines is the vertical projection of each of the 3 visible sides of the box. The longer line corresponds to the side of the box closer to the camera and the shorter lines correspond to the 2 other sides visible along the sequence.

Figure 3 shows a comparison between both versions of EM and the algorithm of Jacobs. The performance of the algorithms was tested as a function of noise level and percentage of missing data. Figure 4 shows result on a real sequence for which EM with temporal coherence succeeded to recover the correct structure, while all other algorithms have failed.

4.2 EM for Motion Segmentation

Figure 5 shows quantitative comparisons of EM and Costeira and Kanade for three different synthetic sequences as a function of noise level. It is apparent that all

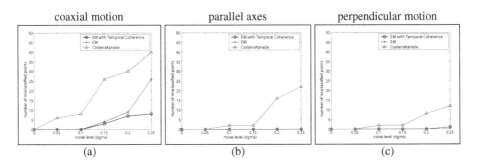

Fig. 5. Comparison of different factorization algorithms for motion segmentation on synthetic inputs. The graphs display total number of misclassified points as a function of the noise standard deviation for $\sigma = 0, \ldots, 0.25$. In some of the experiments, the graphs of the two factor analysis versions overlap. **a.** sequence of concentric cylinders rotating in different speeds. Due to the input degeneracy only EM and [3] are compared. **b.** a cylinder and a cube rotating in the same speed around different parallel axes. **c.** A cube and a cylinder rotating around perpendicular axes.

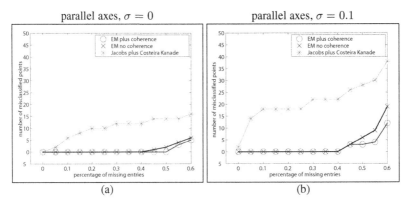

Fig. 6. Performance of EM for motion segmentation with and without temporal coherence. Graphs show number of misclassifications as a function of the percentage of missing data. **a.** Cube and cylinder rotating around different parallel axes without noise. **b.** Cube and cylinder rotating around different parallel axes with noise with standard deviation $\sigma = 0.1$.

algorithms give perfect segmentation when there is no noise at all. As the amount of noise increases, the performance of [3] deteriorates rapidly, while EM-based segmentation continues to succeed for low amounts of noise and shows moderate increase in the number of errors for larger amounts of noise. It is also clear that EM with temporal coherence performs significantly better than EM without temporal coherence for noisy inputs. The algorithms of [5, 17] perform similar to [3] in non-degenerate cases when the actual rank of observation matrix is provided.

Figure 6 shows the performance of EM with temporal coherence as a function of the percentage of missing data. While all other factorization algorithms cannot work with missing data, EM continues to perform well even when 50% of the data is missing. For comparison, we also show the algorithm of [3] when the observation

First Frame	First Object	Second Object

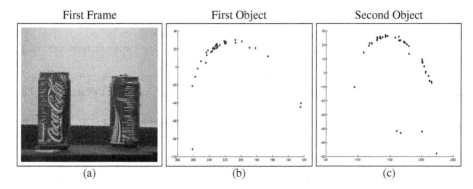

| (a) | (b) | (c) |

Fig. 7. A real sequence of two cans rotating around different parallel axes. EM with temporal coherence succeeds in finding correct segmentation and $3D$ structure reconstruction while other existing algorithms fail. See text for further details. **a.** First image from the input sequence with tracks found by tracking software superimposed. **b.** First segment, top view. **c.** Second segment, top view.

matrix is first filled in using Jacobs' algorithm [10] and the correct rank is given to all algorithms.

Finally, we tested the different algorithms on a real sequence of two cans rotating horizontally around parallel different axes in different angular velocities. 149 feature points were tracked along 20 frames, from which 93 are from one can, and 56 are from the other. Some of the feature points were occluded in part of the sequence, due to the rotation. Notice that despite of its simple appearance, this is a rather challenging scene because a large percentage of the points are missing and because of the motion degeneracy: the two cans have "similar" motion, that is rotation around parallel axes, which leads to a rank deficient motion matrix.

Using EM for motion segmentation, 8 points were misclassified. For comparison, Costeira-Kanade (using the maximal full submatrix of the measurements matrix) resulted in 30 misclassified points and a failure in $3D$ structure reconstruction. Figure 7(a) shows the first frame of the sequence and the tracks superimposed and figures 7(b), 7(c) show the curved surface of the two cylinders recovered correctly.

5   Discussion

In this paper we have presented an EM algorithm for matrix factorization based on representing the factorization problem as a problem of factor analysis.

Working with this representation allowed us to (1) handle correlated measurements noise and missing data, (2) place informative priors on both structure and motion enabling 3D reconstruction in scenes where previous methods have failed and (3) impose constraints on the resulting factors, thereby extending the applicability of factorization methods to problems such as subspace separation.

It would be interesting to study applications of the enhanced factorization capabilities presented in this paper in other vision problems and in problems taken from other areas, for example, semantic analysis of texts ([4]).

References

1. R.A. Andersen and D.C Bradley. Perception of three-dimensional structure from motion. In *Trends in Cognitive Sciences, 2,* pages 222–228, 1998.
2. M.E. Brand. Incremental singular value decomposition of uncertain data with missing values. In *Proceedings of the European Conference on Computer Vision (ECCV),* pages 707–720, May 2002.
3. J. Costeira and T. Kanade. A multi-body factorization method for motion analysis. In *Proceedings of the International Conference on Computer Vision (ICCV),* pages 1071–1076, 1995.
4. Scott C. Deerwester, Susan T. Dumais, Thomas K. Landauer, George W. Furnas, and Richard A. Harshman. Indexing by latent semantic analysis. *Journal of the American Society of Information Science,* 41(6):391–407, 1990.
5. C.W. Gear. Multibody grouping from motion images. *Inernational Journal of Computer Vision (IJCV),* pages 133–150, 1998.
6. A. Gelb, editor. *Applied Optimal Estimation.* MIT Press, 1974.
7. A. Gruber and Y. Weiss. Factorization with uncertainty and missing data: Exploiting temporal coherence. In *Proceedings of Neural Information Processing Systems (NIPS),* 2003.
8. A. Gruber and Y. Weiss. Multibody factorization with uncertainty and missing data using the EM algorithm. In *Proceedings of Computer Vision and Pattern Recognition (CVPR),* 2004.
9. M. Irani and P. Anandan. Factorization with uncertainty. In *Proceedings of the European Conference on Computer Vision (ECCV) (1),* pages 539–553, 2000.
10. D. Jacobs. Linear fitting with missing data: Applications to structure-from-motion and to characterizing intensity images. In *Proceedings of Computer Vision and Pattern Recognition (CVPR),* pages 206–212, 1997.
11. D. D. Morris and T. Kanade. A unified factorization algorithm for points, line segments and planes with uncertain models. In *Proceedings of the International Conference on Computer Vision (ICCV),* pages 696–702, January 1999.
12. S. Roweis. EM algorithms for PCA and SPCA. In *Proceedings of Neural Information Processing Systems (NIPS),* pages 431–437, 1997.
13. D. Rubin and D. Thayer. EM algorithms for ML factor analysis. *Psychometrika 47(1),* pages 69–76, 1982.
14. H. Y. Shum, K. Ikeuchi, and R. Reddy. Principal component analysis with missing data and its application to polyhedral object modeling. *IEEE Transactions on Pattern Analysis and Machine Intellignece (PAMI),* pages 854–867, September 1995.
15. C. Tomasi and T. Kanade. Shape and motion from image streams under orthography: A factorization method. *International Journal of Computer Vision (IJCV),* 9(2):137–154, November 1992.
16. S. Ullman. *The interpertation of visual motion.* MIT Press, 1979.
17. L. Zelnik-Manor, M. Machline, and M. Irani. Multi-body segmentation: Revisiting motion consistency. In *Workshop on Vision and Modeling of Dynamic Scenes, with ECCV,* 2002.

A Simple Feature Extraction for High Dimensional Image Representations*

Christian Savu-Krohn and Peter Auer

Chair of Information Technology (CIT), University of Leoben, Austria
{csavukrohn,auer}@unileoben.ac.at

Abstract. We investigate a method to find local clusters in low dimensional subspaces of high dimensional data, e.g. in high dimensional image descriptions. Using cluster centers instead of the full set of data will speed up the performance of learning algorithms for object recognition, and might also improve performance because overfitting is avoided. Using the Graz01 database, our method outperforms a current standard method for feature extraction from high dimensional image representations.

1 Introduction

One of the key requirements to a modern Cognitive Vision System is a robust performance upon changes in illumination, scale, pose etc. For this, modern feature extraction methods like Lowe's Scale-Invariant-Feature-Transforms [1] and Mikolajczyk's and Schmid's Scale-Invariant-Harris-Laplace- [2] and Affine-Invariant-Interest-Point-Detectors [3] come to hand when learning objects or object categories [4, 5]. Opelt *et al.* [4] used AdaBoost [6, 7] to generate a combination of weak hypotheses, where each weak hypothesis consists of a feature vector with a distance threshold. Since the number of feature vectors is very large, the search for weak hypotheses becomes computationally very expensive. To reduce the computational burden, we want to reduce the number of candidates for weak hypotheses by clustering. This would reduce learning time significantly. Furthermore, this promises less overfitting and, eventually, more accurate classifiers. Note that both, Opelt *et al.* [4] and Dance *et al.* [5] use k-means for that purpose.

Unfortunately, modern descriptors like Lowe's SIFTs reside in high dimensional space for which common metrics like the euclidean might be unsuitable [8]. Therefore, dimensionality reduction techniques such as PCA are commonly applied before clustering. Nevertheless, if the specific clusters reside in various subspaces such global reduction techniques may be inappropriate. Recently, projective clustering methods address this problem by searching for local subspace clusters [9, 10]. Aggarwal *et al.* [9] determine the local subspaces through the smallest eigenvectors of each clusters covariance matrix, whereat the user has to predefine the minimum number of subspace dimensions. Böhm *et al.* [10] propose a density connected clustering algorithm, which searches for

---

* This work was presented in a preliminary version at the First Austrian Cognitive Vision Workshop (ACVW 05), Zell an der Pram, January 2005.

C. Saunders et al. (Eds.): SLSFS 2005, LNCS 3940, pp. 163–172, 2006.

variances below a certain threshold along the attributes to identify subspaces within ϵ-neighborhoods of points. Using another parameter, they limit the number of admissible subspace dimensions from above.

Unfortunately, the number of subspace dimensions is commonly unknown beforehand. Furthermore, the metrics employed by Aggarwal *et al.* [9] and Böhm *et al.* [10] may deteriorate upon high dimensional subspaces. Therefore, we propose a fast projective clustering algorithm (FPC), which aims to find axis-parallel subspace-clusters while determining the number of subspace dimensions automatically. Our approach is to search for the interval of highest density along all coordinate axes recursively [Sec. 2]. Since our actual goal is feature extraction for learning from high dimensional image representations the evaluations are twofold. First, we compare our method to k-means in an unsupervised setting using artificial data [Sec. 3.1]. Then, we evaluate our method within a boosting frame work, enabling us to directly compare our results to those of Opelt *et al.* [4] using k-means [Sec. 3.2].

2   The Clustering Algorithm

Assume that the features reside in $\mathbb{R}^n$. Then the densest interval $[a, b]$ along each coordinate $l \in \{1, ..., n\}$ is calculated. (The details of this calculation are given in the next section.) For the coordinate with the overall densest interval, all data points with a corresponding coordinate in this interval are selected and processed by a recursive application of the algorithm [App. A]. The algorithm terminates if no meaningful dense intervals can be found. The final result of one iteration of the algorithm is a subspace cluster defined by the hyper-rectangle of the recursively chosen coordinates and intervals. When such a cluster is found, the data points in this cluster are removed and the algorithm is restarted for the remaining data. The overall algorithm terminates if no more clusters can be found.

2.1   Calculating the Densest Interval Along a Single Coordinate

Let $\mathbf{x} = (x_1 \leq ... \leq x_m) \in \mathbb{R}$ be an ordered dataset with diameter $r = \max \mathbf{x} - \min \mathbf{x}$. To calculate the densest interval we assume that the data are drawn from a probability density function

$$f(x \mid a, b) = \frac{1-q}{r} + 1_{[a,b]}(x) \frac{q}{b-a} \tag{1}$$

with unknown $[a, b]$ and q, and the indicator function

$$1_{[a,b]}(x) = \left\{ \begin{array}{l} 1, x \in [a, b] \\ 0, else \end{array} \right\} \tag{2}$$

Choosing the parameters which maximize the likelihood of the data we find the desired interval. Optimizing the log-likelihood LL for q we find

$$LL(\mathbf{x} \mid a, b) = (m - \xi) \cdot \log\left(\frac{1 - \frac{\xi}{m}}{r - (b-a)}\right) + \xi \cdot \log\left(\frac{\frac{\xi}{m}}{b-a}\right) \tag{3}$$

with $\xi = |\{x | a \leq x \leq b\}|$. Thus, it remains to select a and b such that $LL(\mathbf{x} | a, b)$ is maximized. Thereby, we restrict ξ as follows.

Viewing clustering as learning an indicator function in an unsupervised setting, we consult the supervised case to infer a reasonable sample bound. Within the agnostic learning framework [11, 12] samples $(\mathbf{x}_1, y_1), ..., (\mathbf{x}_m, y_m)$ are drawn randomly from a joint distribution D over $\mathbb{R}^n \times \{1, ..., \#classes\}$, and the learner's goal is to output a hypothesis $\hat{h} \in \mathcal{H}$ such that for another $(\mathbf{x}, y)$ the probability $P(\hat{h}(\mathbf{x}) \neq y)$ is almost that of the best $h \in \mathcal{H}$:

$$\mathbb{P}\left[P\left(\hat{h}(\mathbf{x}) \neq y\right) - \min_{h \in \mathcal{H}} P(h(\mathbf{x}) \neq y) \leq \epsilon\right] \geq 1 - \delta \tag{4}$$

with $\mathbb{P}$ the probability of drawing $(\mathbf{x}_1, y_1), ..., (\mathbf{x}_m, y_m)$. For this setting it has been shown [13] that one has to sample

$$m \sim \frac{1}{\epsilon^2}(VCdim(\mathcal{H}) - \log \delta) \tag{5}$$

data points. That is, even if labels were given, the bounds $[a, b]$ of the interval remain imprecise by

$$m \cdot \epsilon = m \cdot \sqrt{(VCdim(\mathcal{H}) - \log \delta)/m} \approx \sqrt{m} \tag{6}$$

data points compared to the best interval possible. Therefore, it is reasonable to use

$$s_{min} \leq \xi \leq m - s_{min}, \tag{7}$$

with $s_{min} = \sqrt{m}$, as a validation criterion for a solution $[a, b]$.

This choice of s_{min} guarantees a linear runtime at a maximum level of granularity when maximizing the Likelihood [Eq. 3] using the following exhaustive search procedure. Let

$$\Phi = \left\{\varphi | \varphi = \frac{x_i + x_{i+1}}{2}, 1 \leq i \leq m, x_i \neq x_{i+1}\right\}$$
$$\cup \{\min \mathbf{x}, \max \mathbf{x}\}, \tag{8}$$

where $\mathbf{x} = (x_1 \leq ... \leq x_m) \in \mathbb{R}$ and $\varphi_1 < ... < \varphi_{|\phi|} \in \Phi$, denote the ordered set of possible interval bounds. Since an exhaustive search over Φ would require $O(m^2)$ steps, we use the following heuristic search (which might return a suboptimal interval). In a first phase, we determine a coarse optimal solution by exhaustively searching among the subset $\hat{\Phi} \subseteq \Phi$ of bounds, that pairwise enclose the minimum number s_{min} of points. Therefore, we denote

$$\hat{\Phi} = \{\hat{\varphi}_1 < ... < \hat{\varphi}_{|\hat{\Phi}|}\} \tag{9}$$

as the set of coarse bounds $\hat{\varphi}$ with

$$\hat{\varphi}_1 = \min \Phi = \min(\mathbf{x}),$$
$$\hat{\varphi}_i = \min\{\varphi \mid \varphi > \hat{\varphi}_{i-1}, \#\varphi \geq \#\hat{\varphi}_{i-1} + s_{min}\},$$
$$\#\varphi = |\{x \leq \varphi\}|.$$

That is, we start at the lowest possible bound and iteratively define a new coarse bound after passing s_{min} points. Note, that we allow for the last interval to contain less than s_{min} data points. Using $\hat{\Phi}$, we get a coarse solution by exhaustive search

$$(\hat{\varphi}_i, \hat{\varphi}_j) = \underset{\hat{\varphi}_{i'} < \hat{\varphi}_{j'}}{\arg\max}\ LL(\mathbf{x} \mid \hat{\varphi}_{i'}, \hat{\varphi}_{j'}) \tag{10}$$

Note, that under the initial assumption of ordered data this computation takes $O(m)$. Then, in the second phase, we refine the coarse solution at the granular level. Although the bounds are imprecise according to Equation 6, we are interested in the empirical maximum of the Likelihood [Eq. 3] and, thus, maximize

$$(\varphi_i, \varphi_j) = \underset{(\varphi_{i'}, \varphi_{j'})\ \in\ \Phi^{\hat{\varphi}_i} \times \Phi^{\hat{\varphi}_j}}{\arg\max}\ LL(\mathbf{x} \mid \varphi_{i'}, \varphi_{j'}) \tag{11}$$

using

$$\Phi^{\hat{\varphi}_i} = \{\varphi \mid \hat{\varphi}_{i-1} \le \varphi \le \hat{\varphi}_{i+1}\} \tag{12}$$
$$\Phi^{\hat{\varphi}_j} = \{\varphi \mid \hat{\varphi}_{j-1} \le \varphi \le \hat{\varphi}_{j+1}\}$$

as the set of possible bounds around the coarse bounds. Again, the computation takes $O(m)$ time. Note, that the likelihood for a bimodal distribution is also maximized for (φ_i, φ_j) enclosing the valley between the two modes. Thus, using

$$F(a, b, r, m) = \frac{|\{x : a \le x \le b\}|}{m} \cdot \frac{r}{b - a} \tag{13}$$

we select the densest interval (a, b) to

$$(a, b) = \underset{(a',b')}{\arg\max}\ F(a', b', r, m) \tag{14}$$

with $(a', b') \in \{(\min \mathbf{x}, \varphi_i), (\varphi_i, \varphi_j), (\varphi_j, \max \mathbf{x})\}$. To further refine the selected interval we rerun the algorithm on the data from it until no further valid subinterval [Eq. 7] can be selected. Consequently, at least s_{min} data points are removed within each iteration, which yields a total run time of at most $O(m^{3/2})$.

2.2 Processing High Dimensional Data

Assume the densest interval (a, b) along each coordinate has been calculated [Sec. 2.1] and let Λ denote the set of all coordinates along which there exists a valid refinement of the data [Eq. 7]. Thus, we determine the clusters $\mathbf{C} = \{\mathbf{x} : a_l \le \mathbf{x}_l \le b_l\}$ and $\bar{\mathbf{C}} = \mathbf{X} \setminus \mathbf{C}$ by selecting that coordinate $l \in \Lambda$ which holds the densest interval among all coordinates

$$l = \underset{l' \in \Lambda}{\arg\max}\ F(\mathbf{a}_{l'}, \mathbf{b}_{l'}, \rho_{l'}, m) \tag{15}$$

with ρ the diameter of the full data set. Then, we recurse upon the data in $\mathbf{C}$ by recalculating $(\mathbf{a}, \mathbf{b})$ and Λ until $\Lambda = \{\}$, i.e. there are no more valid coordinates along which cluster $\mathbf{C}$ can be bounded. Consequently, $\mathbf{C}$ is stored, and the algorithm [App. A] restarts upon the remaining data. Finally, the input data is partitioned into a set of subspace clusters, whereat each cluster denotes its constituting bounds in their particular subspace.

3 Evaluation

3.1 Artificial Data

In a first step, we have evaluated our approach on a artificial 25-dimensional dataset containing $k = 19$ axis-parallel clusters $\mathcal{C}$ located in 2-dimensional subspaces. We build the clusters by sampling an increasing number of points $\{100, 150, ..., 1000\}$ from 25-dimensional gaussians $\mathcal{N}(\mathbf{0}, \boldsymbol{\sigma})$ with $\sigma = 1$ except for 2 randomly chosen dimensions with $\sigma = 0.1$. Furthermore, we added 25% of uniformly distributed noise within the range of the data. Let $\hat{\mathcal{C}}$ denote the clustering obtained from a particular clustering method.

We evaluate the quality of a clustering obtained in terms of how well the known clusters are covered. Particularly, we assume that every cluster $\hat{\mathbf{C}}_j$ belongs exactly to one cluster $\mathbf{C}_i$ and, thus, calculate the coverage

$$p = \frac{\sum_j |\mathbf{C}_{\tau(j)} \cap \hat{\mathbf{C}}_j|}{\sum_j |\hat{\mathbf{C}}_j|}$$

$$\tau(j) = \arg \max_i |\mathbf{C}_i \cap \hat{\mathbf{C}}_j|$$

with $i = 1, ..., k$, $j = 1, ..., \hat{k}$ and $\tau = \{1, ..., k\}^{\hat{k}}$.

Unlike k-means, FPC produces a deterministic output upon a certain input. Hence, we compare our clustering of the data to multiple rounds of k-means, each varying in $\hat{k}$ and the seeds sampled from the data at random. Figure 1 shows that FPC outperforms k-means.

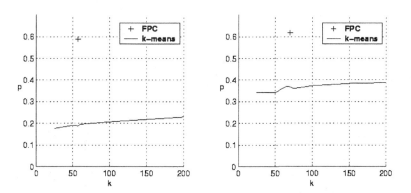

Fig. 1. Coverage of real clusters by the clustering found. Left: at zero noise. Right: at 25% noise level.

3.2 Graz01 Database

Furthermore, we tested our method for image categorization using the Graz01 database[1] using LPBoost [14] as learning method. Particularly, we used the 128-dimensional

---

[1] Available at http://www.emt.tugraz.at/~pinz/data/

SIFTs extracted by Opelt *et al.* [4] from 300 images of the categories 'bike' and 'background', and retained the SIFTs from 50 images per class for testing. Applying FPC on the training set, we obtained 459 subspace clusters from about 400000 SIFTs. It turned out that all clusters were bounded in at most 18 coordinates, indicating the typically

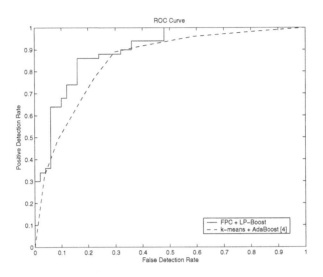

Fig. 2. ROC-Curve of our method compared to Opelt *et al.* [4]

Fig. 3. Detected Features. Upper row: The features with minimum distance to the three weak hypotheses, which have the highest weight (within the ensemble) among all those that triggered for the particular image. Classification is robust to the objects pose. Lower row: **All** features with a distance below the threshold to one of the three weak hypotheses, which have the highest weight (within the ensemble) among all those that triggered for the particular image.

low subspace dimensionality observed. Similar to Opelt *et al.*, we calculate the distance matrix $\mathbf{D}^{k \times \#images}$ of clusters to images before boosting, where each entry denotes the minimum distance between a cluster's center $\mathbf{c}_j$ and the SIFTs from an image. Unlike Opelt *et al.*, we do not consider all SIFTs from an image during the calculation of the distance matrix, but only those that fall into the particular cluster's subspace bounds, setting the distance to infinity if there are no such SIFTs.

Using $\mathbf{D}$, LPBoost calls a weak learner within each boosting round to obtain a optimal weak hypothesis with respect to the current boosting weights. Particularly, a weak hypotheses is derived from a cluster by sorting the distances $\mathbf{D}_{j,.}$ to its center and selecting an optimal threshold θ_j thereupon, such that the sum of the weighted labels from those images with distance below θ_j is maximized. See Opelt *et al.* [4] for details. Using LPBoost [14] we achieve an 86% ROC-equal error rate (with an area of 0.8968 below the curve) [Fig. 2] on the test set, which outperforms Opelt *et al.* [4]. Furthermore, only 28 weak hypotheses, having weights greater zero in the ensemble, contribute to the final hypothesis. Examination of the contributing weak hypotheses showed that our feature extraction focuses on typical structures like bars, tyres, spokes and wires. See figure 3 for some detected features. Particularly, the examples show that classification is robust to variations of the objects pose. Though similar structures may also be detected in the background of images from the positive class or the negative class respectively, the final classification remains correct.

4   Discussion

We have presented a new method for feature extraction from high dimensional image representations. The evaluations approve the viability of our work. Furthermore, generating weak hypotheses from out **FPC** is straightforward since each cluster denotes thresholds in various subspaces of the data. The final ensemble is sparse and outperforms results from earlier work. Hence, overfitting is limited and generalisation performance is improved. Therefore, feature extraction using FPC should be tried out on other applications involving high dimensional data to check if the results translate.

Acknowledgments

This work was supported by the European project LAVA (IST-2001-34405), by the FSP/JRP Cognitive Vision of the Austrian Science Funds (FWF-JRP S9104-N04 SP4) and in part by the IST program of the European Community, under the PASCAL Network of Excellence, IST-2002-506778. This publication only reflects the authors' views.

References

1. Lowe, D.: Object recognition from local scale-invariant features. In: Seventh International Conference on Computer Vision. (1999) 1150–1157
2. Mikolajczyk, K., Schmid, C.: Indexing based on scale invariant interest points. In: Proceedings of the 8th International Conference on Computer Vision, Vancouver, Canada. (2001) 525–531

3. Mikolajczyk, K., Schmid, C.: An affine invariant interest point detector. In: European Conference on Computer Vision. (2002) 128–141

4. Opelt, A., Fussenegger, M., Pinz, A., Auer, P.: Weak hypotheses and boosting for generic object detection and recognition. In: ECCV (2). (2004) 71–84

5. Dance, C., Willamowski, J., Csurka, G., Bray, C.: Categorizing nine visual classes with bags of keypoints. (2004)

6. Freund, Y., Schapire, R.E.: Experiments with a new boosting algorithm. In: International Conference on Machine Learning. (1996) 148–156

7. Freund, Y., Schapire, R.E.: A decision-theoretic generalization of on-line learning and an application to boosting. J. Comput. Syst. Sci. **55** (1997) 119–139

8. Hinneburg, A., Aggarwal, C.C., Keim, D.A.: What is the nearest neighbor in high dimensional spaces? In: The VLDB Journal. (2000) 506–515

9. Aggarwal, C.C., Yu, S.P.: Finding generalized projected clusters in high dimensional spaces. In: SIGMOD '00: Proceedings of the 2000 ACM SIGMOD international conference on Management of data, New York, NY, USA, ACM Press (2000) 70–81

10. Böhm, C., Kailing, K., Kriegel, H.P., Kröger, P.: Density connected clustering with local subspace preferences. In: Proceedings of the 4th IEEE International Conference on Data Mining. (2004) 27–34

11. Haussler, D.: Decision theoretic generalizations of the pac model for neural net and other learning applications. Inf. Comput. **100** (1992) 78–150

12. Kearns, M.J., Schapire, R.E., Sellie, L.: Toward efficient agnostic learning. In: Computational Learing Theory. (1992) 341–352

13. Long, P.M.: The complexity of learning according to two models of a drifting environment. Machine Learning **37** (1999) 337–354

14. Demiriz, A., Bennett, K.P., Shawe-Taylor, J.: Linear programming boosting via column generation. Machine Learning **46** (2002) 225–254

A   The Algorithm

Algorithm 1 - Fast Projective Clustering (FPC)

---

procedure fpc($\mathbf{C}, s_{min}$)
% Input: data set $\mathbf{C} \in \mathbb{R}^{m \times n}$, minimum support s_{min}
% Output: clustering $\mathcal{C}$

  $\rho = \max \mathbf{C} - \min \mathbf{C}$
  while $\mathbf{C}$ is not empty **do**
    repeat
      $[\mathbf{C}, \bar{\mathbf{C}}] = \text{addBounds}(\mathbf{C}, s_{min}, \rho)$
    until $\mathbf{C}$ is not refined
    $\mathcal{C} = \mathcal{C} \cup \{\mathbf{C}\}$
    $\mathbf{C} = \bar{\mathbf{C}}$
  end while

---

procedure selectDensestInterval($\mathbf{C}, \varphi_i, \varphi_j, l$)
% Input: cluster $\mathbf{C} \in \mathbb{R}^{m \times n}$, bounds (φ_i, φ_j), coordinate l
% Output: data $\mathbf{C}$ from densest interval with bounds (a, b), separated data $\bar{\mathbf{C}}$

$r = \max \mathbf{C}_l - \min \mathbf{C}_l$
$\Phi_i \times \Phi_j = \{(\min \mathbf{C}_l, \varphi_i), (\varphi_i, \varphi_j), (\varphi_j, \max \mathbf{C}_l)\}$
$(a, b) = \underset{(a', b') \in \Phi_i \times \Phi_j}{\arg \max} \; F(a', b', r, m)$
$\mathbf{C}, \bar{\mathbf{C}} \overset{(a,b)}{\leftarrow} \mathbf{C}$

---

procedure addBounds($\mathbf{C}, s_{min}, \rho$)
% Input: initial cluster $\mathbf{C} \in \mathbb{R}^{m \times n}$, minimum support s_{min}, diameter ρ of the data set
% Output: refined cluster $\mathbf{C}$ with new bounds (a, b) along coordinate l, separated data $\bar{\mathbf{C}}$

$\Lambda = \{\}$
for $l = 1$ to n **do**
   $\mathbf{C}' = \mathbf{C}$
   $valid = true$
   while $valid$ **do**
      $[\varphi_i, \varphi_j]$ = optimalBounds($\mathbf{C}', s_{min}, l$)
      $[\mathbf{C}', \bar{\mathbf{C}}', a, b]$ = selectDensestInterval($\mathbf{C}', \varphi_i, \varphi_j, l$)
      $valid = (|\mathbf{C}'| \geq s_{min}) \wedge (|\bar{\mathbf{C}}'| \geq s_{min})$
      if $valid$ **then**
         $\Lambda = \Lambda \cup l$
         $\mathbf{a}_l = a$
         $\mathbf{b}_l = b$
      end if
   end while
end for
if Λ is not empty **then**
   $l = \underset{l' \in \Lambda}{\arg \max} \; F(\mathbf{a}_{l'}, \mathbf{b}_{l'}, \rho_{l'}, m)$
   $\mathbf{C}, \bar{\mathbf{C}} \overset{(\mathbf{a}_l, \mathbf{b}_l)}{\leftarrow} \mathbf{C}$
end if

---

procedure optimalBounds($\mathbf{C}, s_{min}, l$)
% Input: cluster $\mathbf{C} \in \mathbb{R}^{m \times n}$, minimum support s_{min}, coordinate l
% Output: optimal bounds φ_i, φ_j maximizing the Likelihood

$\mathbf{x} = sort(\mathbf{C}_l)$
$r = \max \mathbf{x} - \min \mathbf{x}$
if r = 0 **then**
   $\varphi_i = \min \mathbf{x}$
   $\varphi_j = \max \mathbf{x}$
   return
end if
$\Phi = \{\varphi | \varphi = \frac{x_i + x_{i+1}}{2}, 1 \leq i \leq m, x_i \neq x_{i+1}\} \cup \{\min \mathbf{x}, \max \mathbf{x}\}$
$\hat{\Phi} = \{\hat{\varphi}_1 < ... < \hat{\varphi}_{|\hat{\Phi}|}\}$ **with**
   $\hat{\varphi}_1 = \min(\mathbf{x})$,
   $\hat{\varphi}_i = \min \{\varphi \mid \varphi > \hat{\varphi}_{i-1}, \#\varphi \geq \#\hat{\varphi}_{i-1} + s_{min}\}$,
   $\#\varphi = |\{x \leq \varphi\}|$
$(\hat{\varphi}_i, \hat{\varphi}_j) = \underset{\hat{\varphi}_{i'} < \hat{\varphi}_{j'}}{\arg\max} \; LL(\mathbf{x} \mid \hat{\varphi}_{i'}, \hat{\varphi}_{j'})$
$\Phi^{\hat{\varphi}_i} = \{\varphi \mid \hat{\varphi}_{i-1} \leq \varphi \leq \hat{\varphi}_{i+1}\}$
$\Phi^{\hat{\varphi}_j} = \{\varphi \mid \hat{\varphi}_{j-1} \leq \varphi \leq \hat{\varphi}_{j+1}\}$
$(\varphi_i, \varphi_j) = \underset{(\varphi_{i'}, \varphi_{j'}) \in \Phi^{\hat{\varphi}_i} \times \Phi^{\hat{\varphi}_j}}{\arg\max} \; LL(\mathbf{x} \mid \varphi_{i'}, \varphi_{j'})$

Identifying Feature Relevance Using
a Random Forest

Jeremy D. Rogers and Steve R. Gunn

Image, Speech and Intelligent Systems Research Group,
School of Electronics and Computer Science,
University of Southampton, UK

Abstract. It is known that feature selection and feature relevance can benefit the performance and interpretation of machine learning algorithms. Here we consider feature selection within a Random Forest framework. A feature selection technique is introduced that combines hypothesis testing with an approximation to the expected performance of an irrelevant feature during Random Forest construction.

It is demonstrated that the lack of implicit feature selection within Random Forest has an adverse effect on the accuracy and efficiency of the algorithm. It is also shown that irrelevant features can slow the rate of error convergence and a theoretical justification of this effect is given.

1 Introduction

Ensemble algorithms have achieved success in machine learning by combining multiple weak learners to form one strong learner. The Adaboost algorithm [1] and the Bagging algorithm [2] are two examples of this. These methods centre around the idea of diversity in the base learners, which enable good exploration of possible hypotheses. The Random Forest technique [3], exploits this idea by adopting the randomisation principle of [4] to achieve an increase in diversity. The base learners in this algorithm are decision trees and use information gain as the criterion to split each node. These trees usually perform a search through a large number of possible binary splits for every feature in order to find the optimal split for each node. The Random Forest algorithm uses Bagging to generate a training set for each tree. Diversity is injected into the ensemble by choosing a feature randomly at each node in the tree construction and optimising the split over a set of possible split values along that feature. It is demonstrated here that the lack of implicit feature selection within Random Forest can result in a loss of accuracy and efficiency if irrelevant features are not removed, and a theoretical explanation is given for slower convergence.

Due to the random exploration of features, Random Forest lends itself to feature selection well and the measure of feature importance adopted here is the average information gain achieved during forest construction. It is possible to approximate the expected performance of an irrelevant feature when using this measure, [5]. A feature selection technique is introduced here that combines hypothesis testing with this method and it is empirically shown to achieve good algorithm performance and dimensionality reduction.

C. Saunders et al. (Eds.): SLSFS 2005, LNCS 3940, pp. 173–184, 2006.

2   Irrelevant Features and Random Forest

The Random Forest algorithm makes no distinction between the relevance of features during construction of the forest. As the features are selected randomly with equal probability at each node, the performance can suffer significantly from the presence of irrelevant features. Standard decision tree algorithms will select the optimal feature at each split, in terms of maximal information gain. As Random Forest lacks this implicit feature selection, the probability of selecting an irrelevant feature increases with the proportion of irrelevant features present. These irrelevant features can then mislead the algorithm and increase the generalisation error. Also, as irrelevant features are not effective at separating the data, they can result in unnecessarily large trees and therefore, an increased computational load.

An explanation for the effect of irrelevant features on the Random Forest algorithm can be found by considering the space of possible hypotheses. One of the explanations for how ensemble methods work centres around the concept of the margin. The margin for example x, with label y, is the difference between the probability of correct classification and the probability that it belongs to the next most likely class. For binary classification, the margin is simply,

$$marg\,(x, y) = 2P_\theta\,(h\,(x, \theta) = y) - 1, \tag{1}$$

where θ represents the space of possible hypotheses.

The probability of classifying a data point correctly converges to either 1 or 0 depending on the sign of the margin and [3] uses the law of large numbers to prove that the misclassification rate of an ensemble, H, converges asymptotically to the probability over the input space of obtaining an example with a negative margin. However, for a finite number of hypotheses, n, the probability of correct classification is given by the probability that the majority of the ensemble is correct. This is the probability that at least $\lceil n/2 \rceil$ of the base learners predict the correct class,

$$P\,[H\,(x) = y] = \sum_{i=\lceil n/2 \rceil}^{n} (P_\theta\,[h\,(x, \theta) = y])^i\,(1 - P_\theta\,[h\,(x, \theta) = y])^{n-i} \binom{n}{i} \tag{2}$$

As more base hypotheses are added to the ensemble, the error rate converges in line with this function. Therefore, the speed of convergence can be increased by increasing the size of the margin. For Random Forest, the diversity within the base hypotheses is introduced through the combination of bagging, which trains the hypothesis on a random subset of the training data and random input selection [6]. The space of possible base hypotheses is affected by the set of features chosen to represent the data. As a consequence of this, feature selection can alter the margin values of the data. Ideally, this should result in fewer data points having a negative margin and therefore, allow the algorithm to converge to a smaller error rate. But, it also has the ability to increase the size of the margin values and result in faster convergence. Therefore, fewer trees may be needed, which lowers the computational requirement.

3   Identifying Feature Relevance

Many algorithms exist to identify the relevance of different features in terms of how useful they are in predicting the target. Typically, a trade-off is performed between the accuracy of the procedure and the speed of execution. One simple and fast approach to this problem, is to assume that if a feature provides predictive information concerning the target, then a significant degree of correlation will exist between these variables. Therefore, some algorithms calculate some measure of correlation between individual features and the target, to identify relevant features [7].

Some information can be shared between features and it is beneficial if these redundant features are removed, as the information is only required once. Some algorithms record measures of correlation between pairs of features to identify redundancy and alleviate this problem [7], [8], [9].

The measures of correlation can vary between the standard linear correlation, which is limited to only identifying linear correlations in the data, and measures from information theory such as information gain, conditional entropy and symmetrical uncertainty. These correlation-based methods have proved useful for some data, however, there are limitations with these types of technique. Although a significant degree of correlation between variables is indicative that they share information, it does not follow that for variables which share information there will exist a significant degree of correlation. Interaction can occur between features that can mask relevance and redundancy to correlation-based methods. An extreme example of this is the parity (XOR) problem.

Example 1. If the target, Y is given by the exclusive OR of the binary features, X_1 and X_2, then Y is fully described by the features and they are both relevant.

$$Y = X_1 \oplus X_2$$

However, if each feature assumes the values of 1 or 0 with equal probability, then each feature will have a very low degree of correlation to target. This is because knowing the value of one of the features gives no information about the target without knowing the value of the other feature.

In order to account for this interaction, it is necessary to examine feature subsets rather than individual features. This concept is adopted in the definitions of strong and weak relevance, [10], which deem features to be relevant when considered in the context of other features.

3.1   Random Forest for Feature Selection

An effective feature selection algorithm should evaluate subsets of features rather than individuals. The Random Forest algorithm builds a large number of simple classifiers using randomly chosen features and therefore, achieves a good exploration of possible feature subsets. Also, because Bagging is employed as an integral component of the algorithm, not all of the training data is included in the construction of each of the base hypotheses. This *out-of-bag* data then enables evaluation of the feature subsets without the need for an independent test set. For these reasons, Random Forest lends itself to feature selection well.

A method for identifying feature relevance using RF was given [3], by examining an estimate of the test error through the *out-of-bag* data. The estimate of feature importance is obtained through permuting the values of the feature along all of the examples and observing the effect on the estimate of generalisation error. This method can be used for feature selection [11], however, it is suggested [12], that the method is computationally expensive for high dimensional data and that it is necessary to apply a pre-processing technique that examines the individual performance of each feature. However, this pre-processing can eliminate features with strong interaction.

At each node in the construction of a Random Forest, a feature is selected randomly and used to split the node and maximise the information gain. This information gain can be used as a measure of correlation between the feature and the class. These measures of feature importance can be used to increase performance of the learning algorithm [13]. Although these measures appear to be simpler forms of information gain, there are some benefits to using this method over standard information gain. Each terminal node in a decision tree can be viewed as a learner that has been trained on the features that were used in the path from the root. Consequently, the information gain values are not simply measures of the individual feature performance but measures of the ability of the feature in a variety of possible feature subsets. This enables the algorithm to explore the local relevance of each of the features [14], and can allow for relationships within the data. The reliability of this method can be improved by weighting the observed information gain values with a node complexity measure [5]. The advantage of this, as a technique for feature selection, over the random subspace method [6], is that multiple feature subsets can be evaluated within an individual learner, thus yielding a more efficient subset exploration.

4 A Feature Selection Threshold

It is conjectured that the average information gain during the construction of a Random Forest is a measure of feature relevance. As previously discussed, it tests the feature on different areas of the input space and consequently accounts for the different relationships between features. If these measures of feature importance are applied to learning algorithms which are based on decision trees, it also contains the bias of the learning algorithm. In order to use these measures for feature selection, the ability of an irrelevant feature needs to be established. It is not trivial to calculate the expected information gain created by the splitting of a node, when the data is randomly ordered. It depends upon node size and constitution.

Upper and lower bounds for the expected information gain of an irrelevant feature, $E[IG]$, can be established for the splitting of a node of size, n, by considering the different extremes of constitution [5]. This method considers the different arrangements that could result from the data within a given node being projected onto an irrelevant feature. The assumption that is made here is that no data points lie on top of one another and all of the possible split positions are realisable. However, the Random Forest algorithm uses Bagging, which samples the data with replacement to form different sets from which to construct the base learners. As a result of this some data points are selected multiple times and will consequently lie on top of one another. This limits the possible arrangements and results in a higher observed information gain. To account for this, Bagging is performed as usual to form a sample of the training data and then the multiple instances are removed. For the case when the node being split only contains one example of one of the classes, the node constitution is most unbalanced and gives the lower bound,

$$E\left[IG\right] \geq \frac{1}{n} - \frac{n-1}{n}\log_2\frac{n-1}{n} \tag{3}$$

For the case when the node constitution is most balanced and there are equal numbers of each class, an upper bound can be approximated,

$$E\left[IG\right] \leq \left(\frac{n}{2}\right)^{-0.82} \tag{4}$$

The mid-point between these two bounds represents a reasonable estimate of $E\left[IG\right]$ for the splitting of a node of size n.

4.1   Hypothesis Testing

The approximated value $E\left[IG\right]$, is assumed to be the worst case performance of any feature when splitting the node in question. If a feature is irrelevant then the observed value IG, should approximate $E\left[IG\right]$. The measure of feature importance is the average of the observations $\overline{IG}$, and the feature selection threshold is the average of the corresponding values of $E\left[IG\right]$. However, as the technique uses a sample to estimate these values, there is still a good chance that irrelevant features may be chosen. Hypothesis testing can be employed here to discover the degree of confidence of feature relevance. A t-test is adopted here because of its simplicity, although natural alternatives exist which may yield more favourable results. Following the construction of a Random Forest, each feature has a set of observed information gain values and a corresponding set of values that approximate the performance of an irrelevant feature. If the feature is irrelevant, then these values should approximate one another. The variable that is used here is then the difference between these values and is assumed to have a normal distribution.

$$IG_{diff} = IG - E\left[IG\right] \tag{5}$$

The thresholding method is then equivalent to rejecting the feature if the observed mean of this variable, $\overline{IG_{diff}}$, is less than or equal to zero.

A null hypothesis, H_0, can then be set up to represent an irrelevant feature by assuming that the true mean of IG_{diff}, defined as μ, is less than or equal to zero.

$$H_0 : \mu \leq 0 \tag{6}$$

The alternate hypothesis, H_a, must be the complement of this.

$$H_a : \mu > 0 \tag{7}$$

The null hypothesis can then be rejected if the corresponding likelihood is less than some confidence value, γ. If $\overline{IG_{diff}}$ is less than 0 then the null hypothesis cannot be rejected. However, if it is greater than zero then a one tailed t test can be performed, where the null hypothesis can be rejected with confidence $1 - \gamma$ if,

$$\frac{\overline{IG_{diff}} - \mu}{\frac{S}{\sqrt{n}}} > q \tag{8}$$

Where n is the sample size and S is the standard deviation of the sample. The left hand side of this inequality is a variable which has a Student's t distribution with $n - 1$ degrees of freedom. The cumulative density of this distribution at the value q represents the likelihood of the null hypothesis being valid. If this value is less than γ then the null hypothesis is rejected and the feature is deemed to be relevant. Equation 8 simplifies to,

$$\frac{\overline{IG_{diff}} \sqrt{n}}{S} > q \tag{9}$$

It is important to note that while this method will identify features that are relevant with some degree of confidence, the relevance of the remaining features will be unknown. Therefore, as a feature selection technique, this method may discard some relevant features. Another important consideration is that although features are selected if there exists a certain degree of confidence that the feature is relevant, when examining many features some irrelevant features will be selected by chance. The expected number of these is given by the product of the data dimensionality F and the confidence threshold γ.

5   Experiments

5.1   Datasets

Five real data sets are used for these experiments. The Wisconsin Breast Cancer (WBC), Pima Diabetes, Sonar, Ionosphere and Votes are available from the UCI Repository [15].

An artificial dataset called Simple is also created, which consists of 9 features and 300 examples. The output is generated according to the function,

$$Y = X_1^2 + 2X_2 \tag{10}$$

The remaining seven features are irrelevant and consequently this data set should benefit significantly from feature selection algorithms. It is important to note that as the input values to the function are drawn from a uniform distribution on $[0, 1]^9$, feature 2 has a larger influence on the target.

The Friedman dataset, [16] is another artificial dataset data set that is designed for testing feature selection algorithms. It is generated according to the following formula and also contains 5 irrelevant features. It has been thresholded in order to convert it into a binary classification problem and a threshold value of 14 was chosen to yield a reasonably balanced data set.

$$Y = 10\sin(\pi X_1 X_2) + 20\left(X_3 - \frac{1}{2}\right)^2 + 10X_4 + 5X_5 + N(0, 1.0) \tag{11}$$

Another synthetic data set considered here is the Madelon dataset, which was used in the NIPS 2003 Feature Selection Challenge and consists of 2000 examples and 500 features. [17].

5.2   Irrelevant Features and Random Forest

To demonstrate the effect of irrelevant features on Random Forest, the five real data sets are tested with various numbers of additional random features, generated from a uniform distribution. The number of irrelevant features is varied between 0 and 30. For each experiment, the data sets are randomly partitioned into 90% for training and 10% for testing. 100 trees are constructed to form the forest and classify the test data. This is repeated over 100 trials and the results for the extreme cases of 0 and 30 irrelevant features are shown in Table 1.

Table 1. Error rates and average tree sizes in Random forest for 0 and 30 irrelevant features. Values in brackets for the error rates are the corresponding variances.

Data Set	Error(0)	Av. Tree Size(0)	Error(30)	Av. Tree Size(30)
WBC	0.0296(0.0003)	59.0	0.0328(0.0004)	165.0
Pima	0.2487(0.0015)	239.0	0.3004(0.0030)	274.6
Sonar	0.1657(0.0068)	69.6	0.2124(0.0083)	75.5
Ionosphere	0.0856(0.0019)	47.6	0.0942(0.0022)	75.0
Votes	0.0705(0.0013)	50.7	0.1009(0.0017)	122.9

Figure 1 shows how the error rates increase steadily, as more irrelevant features are added to the data and Figure 2 demonstrates how the average tree size also increases with the presence of irrelevant features. It is important to note that in some cases the error rate does not increase as rapidly, because there is sufficient data to allow the trees to grow larger and compensate. This increase in tree size is undesirable as it increases the computational load.

5.3   Feature Selection Thresholding

The expected information gain of an irrelevant feature is approximated at each node of forest construction by the mid-point of the two bounds, given by Equations 3 & 4. This represents the ability of a feature that contains no useful information concerning the target and can be used as a feature selection threshold. Hypothesis testing can also be used as an extension to this by including

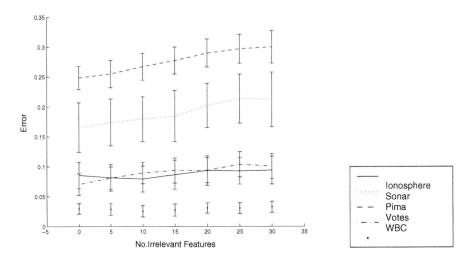

Fig. 1. Error rates of Random Forest on five real data sets with varying numbers of additional irrelevant features. Error bars have a width of one standard deviation recorded over 100 trials.

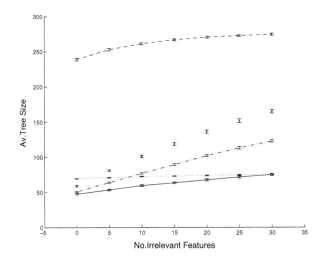

Fig. 2. Average tree sizes created by Random Forest on five real data sets with varying numbers of additional irrelevant features. Error bars have a width of one standard deviation recorded over 100 trials.

only the features that were considered relevant with some degree of confidence. In each experiment, 100 trees are constructed to obtain the average information gain for each feature, using the node complexity measure of [5]. Along with these estimates of feature importance, the corresponding approximations for the performance of an irrelevant feature are also calculated. This information is then used to select a subset of the features and a further 100 trees are constructed based on this subset. Again, the data is partitioned into 90% training and 10% testing and the experiment is repeated over 100 trials. These techniques are compared against RF without feature selection and the CFS algorithm of [9], which is a correlation-based method. The observed error rates are shown in Table 2 and the average number of features selected are shown in Table 3. Table 2 also contains the error rates for the methods when given the previous data sets with 30 additional irrelevant features.

The CFS algorithm performs significantly better on the Votes data set, as this contains a large proportion of redundant features. However, the CFS algorithm significantly degrades performance on a number of data sets as it can eliminate relevant features. It can be seen that using the expected information gain of an irrelevant feature performs well on most data sets and even those where the accuracy is degraded slightly, the dimensionality is greatly reduced.

Table 2. Error rates for RF when using different feature selection strategies. Without feature selection (standard RF), the CFS algorithm (CFS) and using the expected information gain of an irrelevant feature for thresholding (RF Thr) and with hypothesis testing (RF HT).

Data Set	Standard RF	CFS	RF Thr	RF HT
WBC	0.0296(0.0003)	0.0272(0.0003)	0.0243(0.0003)	0.0270(0.0003)
WBC(+30)	0.0328(0.0004)	0.0272(0.0003)	0.0239(0.0003)	0.0225(0.0003)
Pima	0.2487(0.0015)	0.2596(0.0028)	0.2456(0.0023)	0.2388(0.0021)
Pima(+30)	0.3004(0.0030)	0.2596(0.0028)	0.2961(0.0031)	0.2594(0.0026)
Sonar	0.1657(0.0068)	0.2271(0.0068)	0.1638(0.0086)	0.1748(0.0071)
Sonar(+30)	0.2124(0.0083)	0.2271(0.0068)	0.2019(0.0091)	0.1833(0.0070)
Ionosphere	0.0856(0.0019)	0.0581(0.0014)	0.0714(0.0019)	0.0767(0.0011)
Ionosphere(+30)	0.0942(0.0022)	0.0581(0.0014)	0.0772(0.0020)	0.0764(0.0018)
Votes	0.0705(0.0013)	0.0498(0.0010)	0.0650(0.0013)	0.0652(0.0013)
Votes(+30)	0.1009(0.0017)	0.0498(0.0010)	0.0695(0.0012)	0.0684(0.0015)
Friedman	0.1740(0.0075)	0.1795(0.0073)	0.1530(0.0078)	0.1340(0.0049)
Simple	0.0820(0.0024)	0.1917(0.0041)	0.0613(0.0019)	0.0393(0.0014)

Table 3. Proportions of features used with different feature selection strategies. Without feature selection (standard RF), the CFS algorithm (CFS) and using the expected information gain of an irrelevant feature for thresholding (RF Thr) and with hypothesis testing (RF HT).

Data Set	CFS	RF Thr	RF HT
WBC	8.65(96.1%)	9.00(100.0%)	9.00(100.0%)
Pima	3.10(38.8%)	4.00(50.0%)	4.00(50.0%)
Sonar	13.85(23.1%)	58.28(97.1%)	44.20(73.7%)
Ionosphere	14.38(42.3%)	32.79(96.4%)	32.56(95.8%)
Votes	1.00(6.3%)	12.91(80.7%)	12.60(78.8%)
Friedman	3.05(30.5%)	7.99(79.9%)	5.80(58.0%)
Simple	1.00(11.1%)	6.80(75.6%)	4.17(46.3%)

Hypothesis testing is also shown to be beneficial as it enables further dimensionality reduction. For the data sets with the additional irrelevant features, the improvement created by the feature selection techniques over the standard Random Forest is clearly visible. CFS correctly removes all of the irrelevant features as they have no perceivable correlation to the target whilst the Random Forest methods include some of them but manage to remove most.

The accuracy of the RF HT method is dependent upon the number of trees that were constructed to form the estimates of the average information gain for each feature. It is also dependent upon the level of confidence used. Here the Madelon data set [17], is employed as it is already separated into a training and validation set. 10,000 trees are constructed on the training set to form estimates for the average information gain. The confidence level can then be varied to alter the features that are selected. These features are then used to represent the data and a further 1000 trees are constructed on the training set in order to test the validation set. The results are shown in Table 4.

Table 4. Effect of applying feature selection technique to Madelon data set and varying confidence level. The validation set error and the Out-of-Bag estimate of test error (OOB Est.) is shown.

Confidence	Error	OOB Est.	#Features	Av. Tree Size
0.05	0.3550	0.3805	257	920.9
0.025	0.3783	0.3675	232	916.4
0.01	0.3483	0.3620	196	912.3
0.001	0.3483	0.3320	130	898.5
0.0001	0.3300	0.2845	82	876.6
10^{-6}	0.2583	0.2400	52	837.5
10^{-9}	0.1867	0.1700	30	766.0

As the confidence level is lowered, it becomes more difficult to reject the null hypothesis and deem features to be relevant. The Out-of-Bag estimate of test error is gained through testing each training point on the subset of the forest where it was not included in the bagged set. It does not use the test data and can therefore, be used to optimise the confidence level.

5.4   Error Convergence

The effect of feature selection on the error convergence rate can be tested by calculating the error as each tree is added to the forest. This is demonstrated here using the Madelon data set. Two Random Forests are constructed, one using all of the features and one using the features selected by RF HT with a confidence level of 0.05. The results are averaged over 100 trials and shown in Figure 3.

It can clearly be seen that after 100 trees have been added to the forest, the error has not yet converged when all of the features are present. However, when the feature selection scheme is used, the algorithm is very close to convergence.

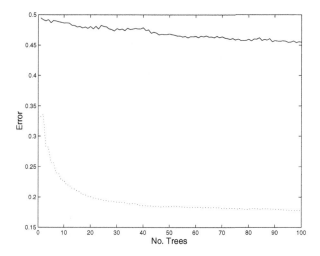

Fig. 3. Comparison of error convergence for Madelon with RF. Solid line shows convergence without feature selection. Dashed line shows convergence with RF HT and confidence level of 0.05.

6 Discussion

The Random Forest algorithm has been shown to suffer significantly here with the presence of irrelevant features. The lack of implicit feature selection within the algorithm allows irrelevant features to be used and consequently, reduces the accuracy. Irrelevant features have also been shown to increase the sizes of the resultant trees and if there is sufficient data the algorithm can compensate, to some extent, by allowing the trees to grow larger. However, larger trees represent an increased computational load and are undesirable. Another reason for the use of feature selection with RF is the effect that irrelevant features have on the error convergence rate. This has been demonstrated here and explained through the concept of the margin.

The average information gain achieved during Random Forest construction achieves good identification of feature relevance by evaluating subsets of features. A feature selection technique has been introduced, which approximates the expected performance of an irrelevant feature when using this measure and employs hypothesis testing to select relevant features with a degree of confidence. This method is shown to work well in terms of improved accuracy and dimensionality reduction.

References

1. Freund, Y., Schapire, R.: A short introduction to boosting. Journal of Japanese Society for Artificial Intelligence **14** (1999) 771–780
2. Breiman, L.: Bagging predictors. Machine Learning **24** (1996) 123–140

3. Breiman, L.: Random forests. Machine Learning **45** (2001) 5–32
4. Dietterich, T.: An experimental comparison of three methods for constructing ensembles of decision trees: Bagging, boosting, and randomization. Machine Learning **40** (2000) 139–157
5. Rogers, J., Gunn, S.: Ensemble algorithms for feature selection. In: The Sheffield Machine Learning Workshop. Volume 3635 of Lecture Notes in Computer Science., Springer (2005)
6. Ho, T.: Nearest neighbours in random subspaces. In: Advances in Pattern Recognition. Volume 1451 of Lecture Notes in Computer Science., Springer (1998) 640–648
7. Roobaert, D., Karakoulas, G., Chawla, N.: Information gain, correlation and support vector machines. In Guyon, I., Gunn, S., Nikravesh, M., Zadeh, L., eds.: Feature Extraction, Foundations and Applications, Springer (2006) In Press.
8. Yu, L., Liu, H.: Feature selection for high-dimensional data: A fast correlation-based filter solution. In: Machine Learning, AAAI (2003) 856–863
9. Hall, M.: Correlation-based feature selection for discrete and numeric class machine learning. In: 17th International Conference on Machine Learning. (2000) 359–366
10. John, G., Kohavi, R., Pfleger, K.: Irrelevant features and the subset selection problem. In Cohen, W., Hirsh, H., eds.: Machine Learning, Morgan Kaufmann (1994) 121–129
11. Svetnik, V., Liaw, A., Tong, C., Wang, T.: Application of breiman's random forest to modeling structure-activity relationships of pharmaceutical molecules. In: In Proceedings of the 5th International Workshop on Multiple Classifier Systems. Volume 3077 of Lecture Notes in Computer Science., Springer (2004) 334–343
12. Chen, Y.W., Lin, C.J.: Combining svms with various feature selection strategies. In Guyon, I., Gunn, S., Nikravesh, M., Zadeh, L., eds.: Feature Extraction, Foundations and Applications, Springer (2006) In Press.
13. Borisov, A., Eruhimov, V., Tuv, E.: Tree-based ensembles with dynamic soft feature selection. In Guyon, I., Gunn, S., Nikravesh, M., Zadeh, L., eds.: Feature Extraction, Foundations and Applications, Springer (2006) In Press.
14. Friedman, J.: Flexible metric nearest neighbour classification (1994)
15. Blake, C., Merz, C.: UCI repository of machine learning databases (1998)
16. Friedman, J.: Multivariate adaptive regression splines. The Annals of Statistics **19** (1991) 1–141
17. Guyon, I., Gunn, S., Nikravesh, M., Zadeh, L., eds.: Feature Extraction, Foundations and Applications. Springer (2006) In Press.

Generalization Bounds for Subspace Selection and Hyperbolic PCA

Andreas Maurer

Adalbertstr. 55, D-80799 München
andreasmaurer@compuserve.com

Abstract. We present a method which uses example pairs of equal or unequal class labels to select a subspace with near optimal metric properties in a kernel-induced Hilbert space. A representation of finite dimensional projections as bounded linear functionals on a space of Hilbert-Schmidt operators leads to PAC-type performance guarantees for the resulting feature maps. The proposed algorithm returns the projection onto the span of the principal eigenvectors of an empirical operator constructed in terms of the example pairs. It can be applied to meta-learning environments and experiments demonstrate an effective transfer of knowledge between different but related learning tasks.

1 Introduction

Humans can use the experience accumulated during previous learning efforts to learn novel but related tasks more efficiently, often generalizing well on the basis of a single training example (see e.g. [13]).

Here we present a machine learning algorithm designed to imitate aspects of this behaviour. It attempts to represent input data in a Euclidean space, such that the metric relations between the represented data points match semantic relations of their class labels. The most elementary semantic relations are equality and inequality, often called *equivalence constraints*, and matching these means that pairs of equally labelled input points should be mapped close to each other, while pairs of differently labelled points should be separated. To train the representing feature map such equivalence constraints can be sampled from environments encompassing many individual learning tasks. If the semantic match is good on the training data and the feature map generalizes well, then we can expect that for any - possibly novel - learning task in the environment, a classifier thresholding the distance to a single training example will have good performance.

Similar methods have received some attention recently, both from the perspective of machine learning ([16],[3]) and cognitive science ([6]). Our approach is motivated by a distribution-independent analysis of the generalization performance of elementary classifiers in a meta-learning environment. The proposed algorithm is a subspace selection technique which can be regarded as a hyperbolic extension of PCA. It utilizes both positive (equal labels) and negative (different labels) equivalence constraints.

C. Saunders et al. (Eds.): SLSFS 2005, LNCS 3940, pp. 185–197, 2006.

The method has been tested in various domains of image recognition: Hand-written characters, rotation and scale invariant character recognition and the recognition of human faces. In all these cases the representations trained on one learning task resulted in a considerable performance improvement for small-sample nearest neighbour classifiers on related tasks.

The next section introduces metric threshold classifiers and corresponding risk functionals for general metric representations. Section 3 presents a probabilistic model for the generation of equivalence constraints. Section 4 specializes to the representations considered by our algorithm and gives a high probability generalization guarantee in terms of the empirical properties of a representation. Section 5 is devoted to the proof of this theorem and section 6 discusses some details of our algorithm. Some experimental results are presented in section 7.

2   Risk Functionals for Metric Representations

Suppose that $\mathcal{E}$ is an environment of learning tasks with common input space $\mathcal{X}$ (see Baxter [4]). This means that $\mathcal{E}$ is a probability distribution on a space of learning tasks $\{(\mathcal{Y}, \mu)\}$, where each $\mathcal{Y}$ is an alphabet of labels, and each μ is a probability distribution on $\mathcal{X} \times \mathcal{Y}$, $\mu(x, y)$ being the probability to encounter the pattern x carrying the label y in the context of the task $(\mathcal{Y}, \mu)$.

We now define a performance measure for metric representations of $\mathcal{X}$ in terms of the expected performance of elementary threshold classifiers. Suppose $\Phi : \mathcal{X} \to \Phi(\mathcal{X})$ is such a representation in a metric space $(\Phi(\mathcal{X}), d)$, where we assume the diameter of $\Phi(\mathcal{X})$ to be bounded by 1. Consider a learning task $(\mathcal{Y}, \mu)$ and a single training example $(x, y) \in \mathcal{X} \times \mathcal{Y}$. A classifier trained on this example alone and applied to another pattern $x' \in \mathcal{X}$ can sensibly only make the decisions "x' is of type y" or "x' is not of type y" or no decision at all. Face recognition is an environment where such classifiers can be quite important in practice: A police officer having to verify the identity of a person on the basis of a single passport photograph has to learn and generalize on the basis of a single example image. A simple classifier using only the metric representation is the threshold classifier $\epsilon_c(x, y)$ which decides

$$\begin{array}{ll}
x' \text{ is of type } y & \text{if } d(\Phi(x), \Phi(x')) < c \\
\text{undecided} & \text{if } d(\Phi(x), \Phi(x')) = c, \\
x' \text{ is not of type } y & \text{if } d(\Phi(x), \Phi(x')) > c
\end{array}$$

where c is some distance threshold $c \in (0, 1)$. Relative to the task $(\mathcal{Y}, \mu)$ this classifier has the error probability (counting 'undecided' as an error)

$$\mathrm{err}\left(\epsilon_c(x, y)\right) = \Pr_{(x', y') \sim \mu} \left\{ r_{\mathcal{Y}}(y, y')(c - d(\Phi(x), \Phi(x'))) \leq 0 \right\},$$

where the function $r_{\mathcal{Y}} : \mathcal{Y} \times \mathcal{Y} \to \{0, 1\}$ quantifies equality and inequality in $\mathcal{Y}$:

$$r_{\mathcal{Y}}(y, y') = \begin{cases} 1 & \text{if } y = y' \\ -1 & \text{if } y \neq y' \end{cases}.$$

The expected value of $\mathrm{err}(\epsilon_c(x,y))$, as a task $(\mathcal{Y},\mu)$ is selected randomly from the environment $\mathcal{E}$ and a training example (x,y) is chosen from μ, is

$$R\left(\Phi,c,\mathcal{E}\right) = \mathbb{E}_{(\mathcal{Y},\mu)\sim\mathcal{E}}\left[\mathbb{E}_{(x,y)\sim\mu}\left[\Pr_{(x',y')\sim\mu}\left\{r_{\mathcal{Y}}\left(y,y'\right)\left(c-d\left(\Phi\left(x\right),\Phi\left(x'\right)\right)\right)\leq 0\right\}\right]\right]$$

The quantity $R\left(\Phi,c,\mathcal{E}\right)$ is a measure of the risk associated with the metric representation Φ, the assumed threshold c and the environment $\mathcal{E}$. Optimization with respect to c gives the threshold independent risk functional[1]

$$R\left(\Phi,\mathcal{E}\right) = \inf_{c\in(0,1)} R\left(\Phi,c,\mathcal{E}\right). \tag{1}$$

Our algorithm will seek a metric representation Φ with a small value of $R\left(\Phi,\mathcal{E}\right)$ where $\Phi\left(\mathcal{X}\right)$ is isometrically embedded in $\mathbb{R}^d$. Any bound on $R\left(\Phi,\mathcal{E}\right)$ is then also a bound on the expected error of threshold classifiers. This is the theoretical justification of the risk functional R, but it does not imply that we are constrained to use the simple and functionally limited threshold classifiers: Any machine learning algorithm applicable to labelled vectors in $\mathbb{R}^d$ (e.g. NN or SVM) can be used on the data which has been preprocessed by Φ.

3   Equivalence Constraints

A triplet $(x,x',r) \in \mathcal{X}^2 \times \{-1,1\}$ is called an *equivalence constraint* ([3],[6]). Given an environment $\mathcal{E}$ we define a probability measure $\rho_{\mathcal{E}}$ on $\mathcal{X}^2 \times \{-1,1\}$ by the formula

$$\rho_{\mathcal{E}}\left(A\right) = \mathbb{E}_{(\mathcal{Y},\mu)\sim\mathcal{E}}\left[\Pr_{((x,y),(x',y'))\sim\mu^2}\left\{(x,x',r_{\mathcal{Y}}\left(y,y'\right)) \in A\right\}\right] \text{ for } A \subseteq \mathcal{X}^2\times\{-1,1\}.$$

To draw an equivalence constraint (x,x',r) from $\rho_{\mathcal{E}}$ we first draw a task $(\mathcal{Y},\mu)$ from $\mathcal{E}$, and then make two independent draws from μ to generate the pair $((x,y),(x',y')) \in (\mathcal{X}\times\mathcal{Y})^2$. If $y=y'$ we set $r=1$ else we set $r=-1$. We then have

$$R\left(\Phi,c,\mathcal{E}\right) = \Pr_{(x,x',r)\sim\rho_{\mathcal{E}}}\left\{r\left(c-d\left(\Phi\left(x\right),\Phi\left(x'\right)\right)\right)\leq 0\right\}. \tag{2}$$

The measure $\rho_{\mathcal{E}}$ is itself unknown to our algorithm, which instead has to rely on a training sample $S = ((x_1,x_1',r_1),...,(x_m,x_m',r_m)) \in \left(\mathcal{X}^2\times\{-1,1\}\right)^m$ of m equivalence constraints generated in m independent, identical trials of $\rho_{\mathcal{E}}$ according to the above procedure, i.e. $S \sim (\rho_{\mathcal{E}})^m$.

The specific way in which the measure $\rho_{\mathcal{E}}$ was generated served to derive and motivate the risk functional R and is otherwise irrelevant to most of our analysis. We only require a probability measure ρ on $\mathcal{X}^2 \times \{-1,1\}$ and risk functionals R as defined by (2) and (1). There are other interesting ways to generate such

---

[1] If 'undecided' was not counted as an error, this infimum would always be attained for some distance threshold $c^* \in [0,1]$, which can be regarded as a granularity of the metric representation.

measures: As pointed out by Bar-Hillel et al ([3]), equivalence constraints can be generated in an unsupervised way by observing a video sequence, regarding image pairs taken at similar times as positive and pairs at very different times as negative constraints. We will therefore axiomatically postulate the existence of the measure ρ, dropping the subscript which indicated the dependence on the environment $\mathcal{E}$. We also write $R(\Phi, c, \mathcal{E}) = R(\Phi, c, \rho_{\mathcal{E}})$ and $R(\Phi, \mathcal{E}) = R(\Phi, \rho_{\mathcal{E}})$.

Another important issue here is balancing. If the alphabets in $\mathcal{E}$ are large, with their symbols appearing approximately equally likely, then negative equivalence constraints will be sampled much more frequently than positive ones, resulting in a negative bias of elementary classifiers. This unwanted effect has been noted in [16] and [3]. A simple remedy is to define a new measure $\bar{\rho}_{\mathcal{E}}$ by

$$\bar{\rho}_{\mathcal{E}}(A) = \frac{\rho_{\mathcal{E}}(A \cap \{1\})}{2\rho_{\mathcal{E}}(X \cap \{1\})} + \frac{\rho_{\mathcal{E}}(A \cap \{-1\})}{2\rho_{\mathcal{E}}(X \cap \{-1\})} \text{ for } A \subseteq \mathcal{X}^2.$$

Then positive and negative equivalence constraints occur equally likely as measured by $\bar{\rho}$. The risk $R(\Phi, c, \bar{\rho}_{\mathcal{E}})$ relative to $\bar{\rho}_{\mathcal{E}}$ is often more relevant than $R(\Phi, c, \rho_{\mathcal{E}})$. Since our bounds will be valid for any probability measure ρ on $\mathcal{X}^2 \times \{-1, 1\}$ they will also work with $\bar{\rho}_{\mathcal{E}}$ as long as we remember that the training sample S is also drawn from the modified measure $S \sim (\bar{\rho}_{\mathcal{E}})^m$.

4   Generalization Bounds for Subspace Selection

Our technique is related to kernel-PCA (see [10], [14]): It requires some fixed map $\psi : \mathcal{X} \to H$ to embed the input data in a Hilbert space H. In practice the embedding ψ is realized by a positive definite kernel κ on the input space which maps onto the inner product $\langle ., . \rangle$ in the Hilbert space H (see [5]). For our results we generally require $\|\psi(x) - \psi(x')\| \le 1$ for all inputs x and x', and we assume H to be infinite dimensional. On the basis of the training set S a d-dimensional orthogonal projection P on H is selected. The combined map of embedding and projection $\Phi = P \circ \psi$ is then used as a metric representation for future data. Since ψ is fixed and P is completely determined by its range, our algorithm can also be considered a *subspace selection technique*.

In the following we fix the Hilbert space H and simply write x instead of $\psi(x)$, identifying $\mathcal{X}$ with its image $\psi(\mathcal{X}) \subset H$ under the kernel-map. When we discuss details of our algorithm we bring ψ back into play. It is crucial that $\text{diam}(\mathcal{X}) \le 1$.

For subspace selection the risk functionals in (2) and (1), which now depend on the projection P, read as

$$R(P, c, \rho) = \Pr_{(x, x', r) \sim \rho} \{ r(c - \|P(x - x')\|) \le 0 \}$$
$$R(P, \rho) = \inf_{c \in (0,1)} R(P, c, \rho).$$

To write down a sample dependent bound on R we introduce for $\gamma > 0$ the margin functions

$$f_\gamma(t) = \begin{cases} 1 & \text{if } t \leq 0 \\ 1 - t/\gamma & \text{if } 0 < t < \gamma \\ 0 & \text{if } \gamma \leq t \end{cases}$$

and the empirical margin error $\hat{R}_\gamma$ for a sample $S = ((x_1, x'_1, r_1), ..., (x_m, x'_m, r_m)) \in (\mathcal{X}^2 \times \{-1, 1\})^m$, a threshold $c > 0$ and a d-dimensional projection P

$$\hat{R}_\gamma(P, c, S) = \frac{1}{m} \sum_{i=1}^m f_\gamma \left(r_i \left(c^2 - \| P(x_i - x'_i) \|^2 \right) \right).$$

Recent results on large margin classifiers (Kolchinskii and Panchenko [7], Bartlett and Mendelson [1]), combined with a reformulation in terms of Hilbert-Schmidt operators give the following:

Theorem 1. *Fix $\gamma > 0$. For every $\delta > 0$ we have with probability greater than $1 - \delta$ in a sample S drawn from ρ^m, that for every d-dimensional projection P*

$$R(P, \rho) \leq \inf_{c \in (0, 1)} \hat{R}_\gamma(P, c, S) + \frac{1}{\sqrt{m}} \left(\frac{2\left(\sqrt{d} + 1\right)}{\gamma} + \sqrt{\frac{\ln(1/\delta)}{2}} \right).$$

In addition to the empirical error the bound shows an estimation error, decreasing as $1/\sqrt{m}$ which is usual for this type of bound. The estimation error contains two terms: The customary dependence on the confidence parameter δ and a complexity penalty consisting of $1/\gamma$ (really the Lipschitz constant of the margin function f_γ), and the penalty $\sqrt{d}$ on the dimension of the representing projection.

We will outline a proof of a more general version of this theorem in the next section.

5   Operator-Valued Linear Large-Margin Classifiers

In this section we rewrite finite dimensional projections and more general feature maps as operator valued large-margin classifiers, and use this formulation to prove a more general version of Theorem 1. We will use the following general result on linear large margin classifiers (Kolchinskii and Panchenko [7], Bartlett and Mendelson [1]):

Theorem 2. *Let (Ω, μ) be a probability space, H a Hilbert space with unit ball $B_1(H)$ and $(w, y) : \Omega \to B_1(H) \times \{-1, 1\}$ a random variable.*
Let $\Lambda \subset H$ be a set of vectors and write

$$B_\Lambda = \sup_{v \in \Lambda} \|v\| \quad \text{and} \quad C_\Lambda = \sup_{v \in \Lambda, \omega \in \Omega} |\langle w(\omega), v \rangle|.$$

Fix $\gamma, \delta \in (0,1)$. Then with probability greater than $1 - \delta$ in $S = (\omega_1, ..., \omega_m)$ drawn from μ^m we have for every $v \in \Lambda$ and every t with $|t| \leq C_\Lambda$

$$\Pr_{\omega \sim \mu} \{y(\omega)(\langle w(\omega), v \rangle - t) \leq 0\}$$

$$\leq \frac{1}{m} \sum_{i=1}^{m} f_\gamma(y(\omega_i)(\langle w(\omega_i), v \rangle - t)) + \frac{1}{\sqrt{m}} \left(\frac{2(B_\Lambda + C_\Lambda)}{\gamma} + \sqrt{\frac{\ln(1/\delta)}{2}} \right).$$

The theorem as stated has an improved margin dependent term by a factor of 2 over the results in [1]. This results from using a slightly different definition of Rademacher complexity with a correspondingly improved bound on the complexity of function classes obtained from compositions with Lipschitz functions (Theorem A6 in [2]).

For a fixed Hilbert space H we now define a second Hilbert space consisting of *Hilbert-Schmidt operators*. With HS we denote the real vector space of symmetric operators on H satisfying $\sum_{i=1}^{\infty} \|Te_i\|^2 \leq \infty$ for every orthonormal basis $(e_i)_{i=1}^{\infty}$ of H. For $S, T \in HS$ and an orthonormal basis (e_i) the series $\sum_i \langle Se_i, Te_i \rangle$ is absolutely summable and independent of the chosen basis. The number $\langle S, T \rangle_{HS} = \sum \langle Se_i, Te_i \rangle$ defines an inner product on HS, making it into a Hilbert space. We denote the corresponding norm with $\|.\|_{HS}$ (see Reed and Simon [12] for background on functional analysis).

We use HS_+ to denote the set of *positive* Hilbert-Schmidt operators,

$$HS_+ = \{T \in HS : \langle Tv, v \rangle \geq 0 \text{ for all } v \in H\}.$$

Then HS_+ is a closed convex cone in HS. Every $T \in HS_+$ has a unique positive squareroot, which is a bounded operator $T^{1/2}$ (in fact $T^{1/2} \in HS_+$) such that $T = T^{1/2}T^{1/2}$.

For every $v \in H$ we define an operator Q_v by $Q_v w = \langle w, v \rangle v$. For $v \neq 0$ chose an orthonormal basis $(e_i)_1^{\infty}$, so that $e_1 = v/\|v\|$. Then

$$\|Q_v\|_{HS}^2 = \sum_i \|Q_v e_i\|^2 = \|Q_v v\|^2 / \|v\|^2 = \|v\|^4,$$

so $Q_v \in HS_+$ and $\|Q_v\|_{HS} = \|v\|^2$. With the same basis we have for any $T \in HS$

$$\langle T, Q_v \rangle_{HS} = \sum_i \langle Te_i, Q_v e_i \rangle = \langle Tv, Q_v v \rangle / \|v\|^2 = \langle Tv, v \rangle.$$

For $T \in HS_+$ we then have

$$\langle T, Q_v \rangle_{HS} = \left\| T^{1/2} v \right\|^2. \tag{3}$$

The set of d-dimensional, orthogonal projections in H is denoted with $\mathcal{P}_d$. We have $\mathcal{P}_d \subset HS_+$ and if $P \in \mathcal{P}_d$ then $\|P\|_{HS} = \sqrt{d}$ and $P^{1/2} = P$.

Consider the feature map given by the operator $T^{1/2}$, where T is any operator in HS_+ (this corresponds to the metric $d(.,.)_T$ considered in [16]). Its threshold dependent risk is

$$R\left(T^{1/2}, c, \rho\right) = \Pr_{(x,x',r)\sim\rho}\left\{r\left(c^2 - \left\|T^{1/2}\left(x - x'\right)\right\|^2\right) \leq 0\right\}$$

$$= \Pr_{(x,x',r)\sim\rho}\left\{r\left(c^2 - \langle T, Q_{x-x'}\rangle_{HS}\right) \leq 0\right\},$$

where we used the key formula (3). For a margin $\gamma > 0$ and a sample $S = ((x_1, x_1', r_1), ..., (x_m, x_m', r_m))$ we define the empirical margin-error

$$\hat{R}_\gamma\left(T^{1/2}, c, S\right) = \frac{1}{m}\sum_{i=1}^m f_\gamma\left(r_i\left(c^2 - \left\|T^{1/2}\left(x_i - x_i'\right)\right\|^2\right)\right)$$

$$= \frac{1}{m}\sum_{i=1}^m f_\gamma\left(r_i\left(c^2 - \langle T, Q_{x_i-x_i'}\rangle_{HS}\right)\right).$$

It is clear that the definitions of R and $\hat{R}$ coincide with those used in Theorem 1 when P is a finite dimensional orthogonal projection. These definitions are also analogous to the risk and empirical margin errors for classifiers obtained by thresholding bounded linear functionals as in Theorem 2. This leads to

Theorem 3. *Let T be some class of positive symmetric linear operators on H and denote[2]*

$$\|T\|_{HS} = \sup_{T\in\mathcal{T}}\|T\|_{HS} \quad and \quad \|T\|_\infty = \sup_{T\in\mathcal{T}}\|T\|_\infty .$$

Fix $\gamma > 0$. Then for every $\delta > 0$ we have with probability greater than $1 - \delta$ in a sample $S \sim \rho^m$, that for every $T \in \mathcal{T}$ and every $c \in \left(0, \|T\|_\infty^{1/2}\right)$

$$R\left(T^{1/2}, c, \rho\right) \leq \hat{R}_\gamma\left(T^{1/2}, c, S\right) + \frac{1}{\sqrt{m}}\left(\frac{2\left(\|T\|_{HS} + \|T\|_\infty\right)}{\gamma} + \sqrt{\frac{\ln(1/\delta)}{2}}\right).$$

Theorem 1 follows immediately from setting $\mathcal{T} = \mathcal{P}_d$, since $\|\mathcal{P}_d\|_{HS} = \sqrt{d}$ and $\|\mathcal{P}_d\|_\infty = 1$.

Proof. Note that for (x, x', r) in the support of ρ we have $\|Q_{x-x'}\|_{HS} = \|x - x'\|^2 \leq 1$, so we can apply Theorem 2 with $\Omega = \mathcal{X}^2\times\{-1,1\}$, $\mu = \rho$, $H = HS$, $w(x, x', r) = -Q_{x-x'}$, and $y(x, x', r) = r$ and $\Lambda = \mathcal{T}$. Then $B_\Lambda = \|T\|_{HS}$ and $C_\Lambda \leq \|T\|_\infty$. Substitution of the expressions for R and $\hat{R}_\gamma$ in the bound of Theorem 2 gives Theorem 3.   □

6   Hyperbolic PCA

Fix a margin γ and a training sample $S = ((x_1, x_1', r_1), ..., (x_m, x_m', r_m))$ of equivalence constraints. Since there are no other sample dependent terms in the bound of Theorem 1, we should in principle minimize the empirical margin-error

$$\hat{R}_\gamma(P, c, S) = \frac{1}{m}\sum_{i=1}^m f_\gamma\left(r_i\left(c^2 - \langle P, Q_{x_i-x_i'}\rangle_{HS}\right)\right).$$

---

[2] Here $\|T\|_\infty = \sup_{\|v\|=1}\|Tv\|$ is the usual operator norm (see [12]).

over all choices of $c \in (0,1)$ and $P \in \mathcal{P}_d$, to obtain a (nearly) optimal projection P^* together with some clustering granularity c^*.

This algorithm is difficult to implement in practice. One obstacle is the non-linearity of the margin functions f_γ. Replacing the f_γ by the convex hinge-loss does not help, because the set of d-dimensional projections itself fails to be convex. Replacing the set $\mathcal{P}_d$ of candidate maps by the set of positive operators with a uniform bound B on their Hilbert-Schmidt norms and replacing the f_γ by any convex function such as the hinge-loss results in a convex optimization problem. Its solution would be the most direct way to exploit Theorem 3 (taking us outside the domain of subspace selection). A major difficulty here is the positivity constraint on the operators chosen. It can be handled by a gradient-descent/projection technique as in [16], but this is computationally expensive, necessitating an eigen-decomposition at every projection step.

Here we take a different path, remaining in the domain of subspace selection. Fix $c \in (0,1)$ and $\gamma > 0$ and for $i \in \{-1,1\}$ define numbers η_i by $\eta_{-1} = \min\left\{\frac{1}{1-c^2}, \frac{1}{\gamma}\right\}$ and $\eta_1 = -\min\left\{\frac{1}{c^2}, \frac{1}{\gamma}\right\}$. Define the empirical operator $\hat{T}(\eta, S)$ by

$$\hat{T}(\eta, S) = \frac{1}{m}\sum_{i=1}^{m}\eta_{r_i}Q_{x_i-x_i'}.$$

Then

$$\hat{R}_\gamma(P,c,S) \leq \frac{1}{m}\sum_{i=1}^{m}\left(1 + \eta_{r_i}\left(c^2 - \langle Q_{x_i-x_i'}, P\rangle_{HS}\right)\right)$$

$$= 1 + \frac{c^2}{m}\sum_{i=1}^{m}\eta_{r_i} - \left\langle\hat{T}(\eta, S), P\right\rangle_{HS}.$$

The right hand side above is the smallest functional dominating $\hat{R}_\gamma$ and affine in the $Q_{x_i-x_i'}$. Minimizing it over $P \in \mathcal{P}_d$ is equivalent to maximizing $\left\langle\hat{T}(\eta, S), P\right\rangle_{HS}$ and constitutes the core step of our algorithm where it is used to generate candidate pairs (P, c) to be tried in the bound of Theorem 1, leading to a heuristic minimization of $\hat{R}_\gamma(P, c, S)$ for different values of $c \in (0,1)$. Current work seeks to replace this heuristic by a more systematic boosting scheme.

Maximization of $\left\langle\hat{T}(\eta, S), P\right\rangle_{HS}$ is carried out by solving the eigenvalue problem for $\hat{T}$ and taking for P the projection onto the span of the d eigenvectors corresponding to the largest eigenvalues of $\hat{T}$. This is similar to the situation for PCA, where the empirical operator approximating the covariance operator is

$$\hat{C}(S) = \frac{1}{m}\sum_{i=1}^{m}Q_{x_i}$$

and the x_i are the points of an unlabeled sample. The essential difference to PCA is that while $\hat{C}$ is a positive operator, the operator $\hat{T}$ is not, and it can have negative eigenvalues. The infinite dimensionality of H and the finite rank of

$\hat{T}$ ensure that there is a sufficient supply of eigenvectors with nonnegative eigenvalues. Nevertheless, while the level sets of the quadratic form defined by $\hat{C}$ are always ellipsoids, those of $\hat{T}$ are hyperboloids in general, due to the contributions of positive equivalence constraints. In a world of acronyms our algorithm should therefore be called HPCA, for *hyperbolic principal component analysis*. If there are only negative equivalence constraints our algorithm is essentially equivalent to PCA.

To describe in more detail how the method works, we put the kernel map ψ back into the formulation. The empirical operator then reads

$$\hat{T}(\eta, S) = \frac{1}{m} \sum_{i=1}^{m} \eta_{r_i} Q_{\psi(x_i) - \psi(x'_i)}.$$

Clearly an eigenvector w of $\hat{T}$ must be in the span of the $\{\psi(x_i) - \psi(x'_i)\}_{i=1}^{m}$, so we can write

$$w = \sum_{i=1}^{m} \alpha_i (\psi(x_i) - \psi(x'_i)). \tag{4}$$

Substitution in the equation $\hat{T}(\eta, S) w = \lambda w$ and taking inner product with $(\psi(x_j) - \psi(x'_j))$ gives the generalized matrix-eigenvalue problem

$$\Gamma D \Gamma \alpha = \lambda \Gamma \alpha$$
$$\Gamma_{ij} = \langle \psi(x_i) - \psi(x'_i), \psi(x_j) - \psi(x'_j) \rangle$$
$$D_{ij} = \eta_{r_i} \delta_{ij}.$$

Evidently all these quantities can be computed from the kernel matrix $\langle \psi(x_i), \psi(x_j) \rangle$. The d solutions $\alpha_k = (\alpha_i)_k$ corresponding to the largest eigenvalues λ are substituted in (4), the resulting vectors w_k are normalized and the projection corresponding to largest eigenvalues is computed. Notice how this algorithm resembles PCA if there are only negative equivalence constraints, because then D becomes the identity matrix.

There is an interesting variant of this method, which is useful in practice even though it does not completely fit the probabilistic framework described above. Suppose we are given an ordinary sample of labelled data $S = ((x_1, y_1), ..., (x_m, y_m))$ and we want to exploit *all* the equivalence constraints implied by S, that is to maximize $\left\langle \hat{T}(\eta, S^{(2)}), P \right\rangle_{HS}$ with $S^{(2)} = ((x_i, x_j, r(y_i, y_j)))_{i \neq j}$. One might be led to think that this would require solving the eigenvalue problem of an $m^2 \times m^2$-matrix, which would be of order m^6, making it computationally impractical even for moderate sample sizes. The problem may however be reduced to the eigenvalue problem of an $m \times m$-matrix, thus of order m^3:

The empirical operator now reads (with $r_{ij} = r(y_i, y_j)$)

$$\hat{T}\left(\eta, S^{(2)}\right) = \frac{1}{m^2} \sum_{i,j=1}^{m} \eta_{r_{ij}} Q_{\psi(x_i) - \psi(x_j)}.$$

Substituting an eigenvector $w = \sum_1^m \gamma_k \psi(x_k)$ in the eigenvalue-equation and taking the inner product with some $\psi(x_l)$, and using the fact that the matrix $a_{ij} = \eta_{r_{ij}}/m^2$ is symmetric we get

$$\lambda \sum_{k=1}^m \gamma_k \langle \psi(x_k), \psi(x_l) \rangle$$

$$= \frac{1}{m^2} \sum_{i,j=1}^m \eta_{r_{ij}} \langle Q_{\psi(x_i) - \psi(x_j)} w, \psi(x_l) \rangle$$

$$= \sum_{k=1}^m \gamma_k \sum_{i,j=1}^m a_{ij} \langle \psi(x_k), \psi(x_i) - \psi(x_j) \rangle \langle \psi(x_i) - \psi(x_j), \psi(x_l) \rangle$$

$$= 2 \sum_{k=1}^m \gamma_k \sum_{ij=1}^m \left(\delta_{ij} \sum_{n=1}^m a_{in} - a_{ij} \right) \langle \psi(x_k), \psi(x_i) \rangle \langle \psi(x_j), \psi(x_l) \rangle .$$

Using G to denote the ordinary Gramian or kernel-matrix $G_{ij} = \langle \psi(x_i), \psi(x_j) \rangle$ we again obtain a generalized $m \times m$ eigenvalue problem

$$GAG\gamma = \lambda G\gamma.$$

A is not diagonal in this case, but given by the symmetric matrix

$$A_{ij} = 2 \left(\delta_{ij} \sum_{n=1}^m a_{in} - a_{ij} \right) = \frac{2}{m^2} \left(\delta_{ij} \sum_{n=1}^m \eta_{r_{in}} - \eta_{r_{ij}} \right) .$$

The sample $S^{(2)}$ does not fit into our probabilistic framework, because it has not been generated by m^2 *independent* draws of equivalence constraints, in fact only $O(m)$ of the pairs in S' can be independent. We nevertheless used this variant of the algorithm to exploit all the information in the training samples for the experiments reported below. The worst possible effect of the use of $S^{(2)}$ is that the number m^2 of equivalence constraints must be replaced by m in our bounds.

7   Experiments

The experiments are designed to test the transfer capabilities of our subspace selection algorithm: We use the data of one set of learning tasks to train a projection, and then check how it facilitates the learning of a new and unknown task.

In practice we take a sample S from a single multiclass learning task with alphabet $\mathcal{Y}$ (this could easily be extended to a collection of tasks) and employ the algorithm described at the end of the previous section to generate projections from all the equivalence constraints implied by S for different values η_1 and η_{-1}, selecting the projection P^* giving the smallest empirical risk $\hat{R}_{0.01}(P^*, S^{(2)})$. The optimal values are reported below for each experiment[3]. Here the balanced version of the risk is used to eliminate the effects of alphabet-sizes.

---

[3] Theorem 1 overestimates the estimation error. This is why a small value for γ is chosen, even though this may make the bound of Theorem 1 trivial.

The projection P^* is applied to a *target task* (with alphabet $\mathcal{Y}'$) for which a test-sample S' is available. The empirical distribution of S' is used to estimate the balanced risk R^* of P^* in the new task (reported below for each case, together with the optimal distance threshold c^*). In addition the feature map is tested with nearest neighbour classification: From S' a *single* example per class is chosen as training data for a nearest neighbour classifier and the error rate of this classifier is recorded for both the metric induced by the feature map (projected data) and the original Euclidean metric on normalized pixel vectors (raw data). This experiment is repeated over all possible choices of training data (in the manner of a leave-$(n-1)$-out test) and the resulting error rates are reported.

The pixel vectors were normalized to unit length. The raw data below already refers to these unit vectors. The embedding ψ was realized by the RBF-kernel $\kappa\left(x,y\right) = 2^{-1}\exp\left(-C\left|x-y\right|^2\right)$, with $C = 16$ for the handwritten digits and $C = 8$ in all other cases [4]. Note that the normalization of the kernel is chosen to bound the diameter of the embedded input vectors by 1, as required by our bounds.

We tried five learning environments, two realistic ones involving handwritten characters and face recognition, and three slightly artificial ones defined by the respective invariances of rotation, scaling and combined rotation and scaling.

For handwritten characters we used images of upper and lower case *letters* in the NIST database to train P^*, and a subset of the MNIST database of *digits* for testing. For face recognition we used the images of 31 subjects in the AT&T Face-Database for training and the remaining 9 subjects for testing.

For rotation invariant character recognition randomly rotated images of printed lower case letters were used for training, randomly rotated images of printed digits (with '9' omitted) for testing. For scale invariant character recognition randomly scaled (from 50% to 150%) images of printed capitals and lowercase letters were used for training, randomly scaled images of printed digits for testing. For combined rotation and scale invariant character recognition the images in the rotation invariant dataset were also randomly scaled (from 50% to 150%). Again the projection was trained from the letters and tested with digits. The following table summarizes the results of these experiments:

The classification error on the projected data correlates well with the risk R^* and the projection leads to a significant improvement in all cases, handwritten character recognition being the most difficult environment. In the case of face recognition the data set used to train the projection is rather small and further improvements are to be expected for larger, perhaps more difficult data sets than AT&T. In the cases, where the environment corresponds to a class of specific geometric invariances, the projection spectacularly reduces the classification error by orders of magnitude.

It seems promising to extend these experiments to the recognition of spatially rotated objects. A very interesting possible line of possible experiments involves unsupervised learning through the observation of a continuous process.

---

[4] Here and in the definition of the kernel $|.|$ refers to the euclidean norm of the pixel vectors.

Table 1. Summary of experimental results

	handw. chars	faces	rotated chars	scaled chars	rotated +scaled chars
$|\mathcal{Y}|$ (Training task)	52	31	20	44	20
$|S|$	4160	310	2000	1320	4000
$|\mathcal{Y}'|$ (Testing task)	10	9	9	10	9
$|S'|$	500	90	900	300	1800
$d = \dim(P^*)$	24	20	18	24	18
η_{-1} ($\eta_1 = 1$, balanced)	0.052	0.016	0.019	0.22	0.19
R^* balanced	0.188	0.05	0.022	0.02	0.068
c^* (balanced)	0.26	0.45	0.3	0.36	0.25
1-NNError on raw data	0.549	0.116	0.716	0.472	0.803
1-NNError on projected data	0.318	0.043	0.014	0.008	0.072

A pair consisting of the present observable vector and a recent memory would be treated as a positive equivalence constraint, a pair of the current vector and a distant memory a negative one. A correspondingly trained projection should map temporal proximity to spatial proximity in its feature space. The observation of continuously and quickly rotating objects which are occasionally being replaced could then lead to a nearly rotation invariant preprocessor. Some experiments pointing in a similar direction have been made by Bar-Hillel et al [3].

References

1. P. L. Bartlett and S. Mendelson. Rademacher and Gaussian Complexities: Risk Bounds and Structural Results. *Journal of Machine Learning Research*, 2002.
2. P.Bartlett, O.Bousquet and S.Mendelson. Local Rademacher complexities. Available online: http://www.stat.berkeley.edu/~bartlett/papers/bbm-lrc-02b.pdf.
3. A. Bar-Hillel, T. Hertz, N. Shental, D. Weinshall. Learning a Mahalanobis Metric from Equivalence Constraints. *Journal of Machine Learning Research* 6: 937-965, 2005.
4. J.Baxter, A Model of Inductive Bias Learning, *Journal of Artificial Intelligence Research* 12: 149-198, 2000
5. Nello Cristianini and John Shawe-Taylor, Support Vector Machines, *Cambridge University Press*, 2000.
6. R. Hammer, T. Hertz, S. Hochstein, D. Weinshall. Category learning from equivalence constraints. XXVII Conference of Cognitive Science Society (CogSci2005), available online.
7. V. Koltchinskii and D. Panchenko, Empirical margin distributions and bounding the generalization error of combined classifiers, *The Annals of Statistics*, Vol. 30, No 1, 1-50.
8. M. Ledoux and M. Talagrand, *Probability in Banach Spaces: isoperimetry and processes.* Springer, 1991.
9. Colin McDiarmid, Concentration, in *Probabilistic Methods of Algorithmic Discrete Mathematics*, p. 195-248. Springer, Berlin, 1998.

10. S.Mika, B.Schölkopf, A.Smola, K.-R.Müller, M.Scholz and G.Rätsch. Kernel PCA and De-noising in Feature Spaces, in *Advances in Neural Information Processing Systems* 11, 1998.

11. J. Shawe-Taylor, N. Christianini, Estimating the moments of a random vector, *Proceedings of GRETSI 2003 Conference*, I: 47–52, 2003.

12. Michael Reed and Barry Simon. *Functional Analysis*, part I of *Methods of Mathematical Physics, Academic Press*, 1980.

13. A. Robins, Transfer in Cognition, in *Learning to Learn*, S. Thrun, L. Pratt Eds. Springer 1998.

14. John Shawe-Taylor, Christopher K. I. Williams, Nello Cristianini, Jaz S. Kandola: On the eigenspectrum of the gram matrix and the generalization error of kernel-PCA. *IEEE Transactions on Information Theory* 51(7): 2510-2522, 2005

15. S.Thrun, Lifelong Learning Algorithms, in *Learning to Learn*, S.Thrun, L.Pratt Eds. Springer 1998

16. E. P. Xing, A. Y. Ng, M. I. Jordan, S. Russel. Distance metric learning, with application to clustering with side information. In S. Becker, S. Thrun, K. Obermayer, eds, *Advances in Neural Information Processing Systems* 14, Cambridge, MA, 2002. MIT Press.

Less Biased Measurement of Feature Selection Benefits

Juha Reunanen

ABB, Web Imaging Systems, P.O. Box 94, 00381 Helsinki, Finland
Juha.Reunanen@iki.fi

Abstract. In feature selection, classification accuracy typically needs to be estimated in order to guide the search towards the useful subsets. It has earlier been shown [1] that such estimates should not be used directly to determine the optimal subset size, or the benefits due to choosing the optimal set. The reason is a phenomenon called overfitting, thanks to which these estimates tend to be biased. Previously, an outer loop of cross-validation has been suggested for fighting this problem. However, this paper points out that a straightforward implementation of such an approach still gives biased estimates for the increase in accuracy that could be obtained by selecting the best-performing subset. In addition, two methods are suggested that are able to circumvent this problem and give virtually unbiased results without adding almost any computational overhead.

1 Introduction

Feature selection is the art of choosing a small yet descriptive subset of useful features from amongst a larger set of candidate features. There may be many reasons for doing this: one might, for example, wish to gain a deeper understanding of the prediction problem at hand, or simply to avoid the potentially costly measurement of all the features. Whatever the aim, it makes sense to assume that one should be able to identify the optimal feature subset size. Moreover, it is often desirable to have the possibility to estimate the increase in accuracy due to choosing an optimally sized subset instead of using all the candidate features.

2 Background

This section briefly describes some basic components required in a feature selection process: the classifier architecture, the evaluation mechanism for a single subset, and the search strategy for finding the useful subsets.

2.1 Classification

A plethora of approaches have been suggested for building automatic feature-based classifiers [see, e.g., 2]. In the context of this paper, the choice of the classifier architecture should be largely irrelevant. However, to verify the generality of the results, they are computed using two very different methods: the 1 nearest neighbor (1NN) classification rule [see, e.g., 3], which is quite popular in feature selection literature, and the C4.5 decision tree building algorithm [4].

C. Saunders et al. (Eds.): SLSFS 2005, LNCS 3940, pp. 198–208, 2006.

2.2   Cross-Validation

In order to guide the search for the optimal subset, a mechanism for determining the potential performance of a single subset is needed. This paper positions itself in the context of the *wrapper* approach [5, 6], where the subsets are evaluated using actual classifiers. A common choice is cross-validation (CV) [7], where the data available is first split into a number of folds. Then, one fold at a time is designated as the validation set, while the others are used for training. The validation set is classified using the classifier that is trained with the corresponding training set. When the errors for the different validation sets are counted up, an estimate for the classification performance using a specific subset is obtained.

The special case when the number of folds is equal to the number samples is usually called "leave-one-out cross-validation" (LOOCV).

Often, it is beneficial to retain the proportions of the different classes between all the folds. If this is enforced, the CV process is called *stratified*. As stratification is known to improve the accuracy of cross-validation [8], it is done in all the experiments of this paper.

2.3   Search Algorithms

Out of the myriad of algorithms suggested for the order in which the feature subsets should be evaluated, this paper experiments with two: Sequential Forward Selection, or SFS [9], and Sequential Floating Forward Selection, SFFS [10]. Both start the search with an empty subset, and, during one iteration, consider the insertion of each feature that still remains excluded. Out of these, the one whose addition results in the largest increase (or smallest decrease) in estimated performance is added to the current set. The difference between the algorithms is that SFFS allows backtracking during the search: after adding a feature, each feature currently selected is subjected to removal. The candidate most promising for deletion is pruned, if doing so yields a better performing subset of the corresponding size than was found previously.[1] In the experiments of this paper, the search is carried on until all the candidate features have been included. This way, the algorithms are able to propose a subset for each possible subset size.

2.4   Interpretation of the Results

Once the search algorithm together with the subset evaluation method has suggested several subsets of different sizes, the practitioner obviously wants to know how these subsets compare to each other: which subset size is the optimal one, and how much better is the optimal subset of that size compared to the full set containing all the candidate features? Answering these two questions is the essence of this paper.

3   The Problem

During the search process, the subset evaluation method, such as cross-validation, produces estimates that are used primarily to guide the search. However, these intermediate

---

[1] The bug fix pointed out by Somol et al. [11] is utilized in this paper.

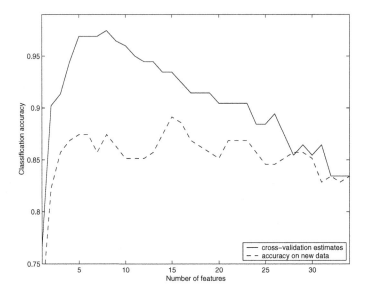

Fig. 1. Observed classification accuracies for the subsets of different sizes found by running a feature selection algorithm, as estimated with LOOCV during the search (solid line) and calculated for independent test data not seen during the search (dashed line). The data is from a single run of the experiments summarized in Table 2.

results can also be stored for later use. Once the search has finished, one could then use the same numbers to compare to each other the subsets of different sizes. An example curve drawn based on such values is shown in Fig. 1 (the solid line).

Unfortunately, when a number of such scores due to cross-validation are compared to each other to facilitate the identification of the best one, *the estimate for the winner of the comparison is no more an unbiased estimate for the accuracy of the winner.* This perhaps counterintuitive fact has been shown many times; from a pedagogical point of view, one of the most successful explanations was given by Jensen and Cohen [12].

In the context of an algorithm like SFS or SFFS, the winner subset for each size has been picked using the cross-validation values required to guide the search. This selection process renders the CV estimate for the winner of each cardinality largely useless for any later comparisons [1]. As a matter of fact, the accuracy obtained for new data tends to behave rather differently, as is shown by the dashed line of Fig. 1.

4   Existing Solutions

This section describes two methods representative of the current state of the art in determining the optimal subset size, and the performance of the best subset of that size.

4.1   Independent Test Set

It is straightforward to use an independent test set to evaluate the performances of the newly found subsets, if such a set happens to be available. On the other hand, if one has

1.	Choose a feature selection algorithm, such as SFS or SFFS.
2.	Divide all the data available into K folds.
3.	**for** $k := 1$ **to** K, (The evaluation loop.)
3.1	Create a (training) set $T^{(-k)}$, which includes the samples of all the folds, except those in the kth.
3.2	Using $T^{(-k)}$, perform a single run of the feature selection algorithm. An inner loop of cross-validation may divide $T^{(-k)}$ further.
3.3	Train classifiers using $T^{(-k)}$ and the obtained subsets of different sizes.
3.4	Test these classifiers using the samples in the kth fold, and record the performance. In the text, these estimates are referred to as $x_i^{(k)}$, $i = 1, 2, \ldots, D$, where D is the total number of candidate variables.
	end;
4.	For each subset size, determine the average of the K estimates obtained in step 3.4: $\bar{x}_i = \frac{1}{K} \sum_{k=1}^{K} x_i^{(k)}$.
5.	Out of all the subset sizes, find the one which maximizes the average score: $\hat{d} = \arg\max_i(\bar{x}_i)$. This is the estimate for the optimal subset size.
6.	The corresponding mean value, $\hat{x} = \max_i(\bar{x}_i)$, is the estimate for the maximum performance that can be attained.
7.	Perform a single run of the feature selection algorithm having all the data available.
8.	Choose the winner amongst the subsets having the size defined in step 5.

Algorithm 1. Determining the optimal subset size and the corresponding performance using an outer loop of cross-validation

1–3.	Perform steps 1–3 of Algorithm 1, including the substeps.
4.	**for** $k := 1$ **to** K, (The indexing loop.)
4.1	For each subset size, take the average of the $K - 1$ estimates obtained for the *other* folds: $\bar{x}_i^{(-k)} = \frac{1}{K-1} \left(\sum_{j=1}^{k-1} x_i^{(j)} + \sum_{j=k+1}^{K} x_i^{(j)} \right)$.
4.2	Find the subset size using which maximum performance is attained: $\hat{d}^{(-k)} = \arg\max_i \left(\bar{x}_i^{(-k)} \right)$.
4.3	Record the performance for the best subset of size $\hat{d}^{(-k)}$ that was obtained during the kth iteration: $x_{\hat{d}^{(-k)}}^{(k)}$.
4.4	For comparison purposes, you can also record the performance for the *full* feature set on the kth iteration: $x_D^{(k)}$.
	end;
5.	Average all the K subset sizes obtained during the different executions of step 4.2. This average is the estimate for the optimal subset size.
6.	Average all the K performance estimates obtained in step 4.3. This average is the estimate for the performance of the best subset having the size discovered during step 5.
7–8.	Perform steps 7–8 of Algorithm 1.

Algorithm 2. The cross-indexing A algorithm. Median or other statistical descriptors could also be used instead of the average in steps 4.1, 5 and 6, if applicable.

access to more data, then one usually wants to append it to the previous dataset, in order to maximally benefit from it. In Sect. 6 of this paper, independent test sets are used to provide the ground truth, using which the approaches being tested can be compared.

4.2 Outer Loop of Cross-Validation

It has been mentioned before that an outer loop of cross-validation could be used to determine the performance of the different subsets, and hence to facilitate the choice of the optimal feature subset [1]. For example, this is what could happen: Using an outer CV loop, a researcher obtains the properly cross-validated estimates for the performance of each subset size. Then, the researcher compares the largest of these values to the estimated performance of the full set, and finds an increase of several percentage points in classification accuracy. This approach is detailed in Algorithm 1.

However, with dozens of candidate features, this method leads to overfitting on yet another level. This is because — once again — the researcher is first determining the maximum of a number of estimates, and then using that estimate.

The estimated performance for the best subset of size i found during iteration k can be thought of as a random variable $X_i^{(k)}$, realizations of which are denoted by $x_i^{(k)}$ in Algorithm 1. The random variable representing the estimate for the performance of the optimal subset (of the optimal size) is

$$\hat{X} = \max_i(\bar{X}_i) = \max_i(\frac{1}{K}\sum_{k=1}^{K} X_i^{(k)}).$$

A lengthy proof given by Jensen and Cohen [12, pp. 318–320] can be used almost as such to show that if every $\bar{X}_i$ is an unbiased estimate for the corresponding true performance ψ_i of a classifier built using the selected feature subset of size i, and there exists no subset size that is the optimal one in all the possible outcomes of the algorithm, then $\hat{X} = \max_i(\bar{X}_i)$ is a positively biased estimator of any ψ_i. This, in turn, makes it a positively biased estimator of the performance attainable with the best optimally sized feature subset.

5 Cross-Indexing

To obtain a truly unbiased estimate, one shall not use *any* such (probabilistic) value that was previously used to pick a certain model, or subset, from a large set of candidates. This principle is reflected in the two algorithms — called *cross-indexing A and B* — that this paper suggests for estimating the performance of the optimal feature subset.

Let us first discuss approach A, delineated in Algorithm 2. In step 4.1, averaged estimates for each subset size are computed much like in Algorithm 1 (step 4) but separately for each fold k, always ignoring the results obtained during the kth iteration of the evaluation loop. These estimates are then used in step 4.2 to determine the optimal subset size $\hat{d}^{(-k)}$. The corresponding performance estimate is obtained by recalling the performance estimated during iteration k of the evaluation loop for the subset having size $\hat{d}^{(-k)}$ (step 4.3). In the end, the results for the k iterations are averaged to produce the final estimates (steps 5 and 6). Next, it is shown that the positive bias of this approach, if any, is bounded by that of outer-loop CV.

Proposition 1. *The estimate $\hat{x}_A$ provided by Algorithm 2 is not more optimistic than the $\hat{x}$ obtained using Algorithm 1.*

Proof. The estimates provided by Algorithms 1 and 2 can be written as follows:

$$\hat{x} = \max_i \left(\frac{1}{K} \sum_{k=1}^{K} x_i^{(k)} \right) = \frac{1}{K} \sum_{k=1}^{K} x_{\hat{d}}^{(k)}, \quad \text{where}$$

$$\hat{d} = \arg\max_i \left(\frac{1}{K} \sum_{\ell=1}^{K} x_i^{(\ell)} \right) = \arg\max_i \left(\sum_{\ell} x_i^{(\ell)} \right), \quad \text{and}$$

$$\hat{x}_A = \frac{1}{K} \sum_{k=1}^{K} x_{\hat{d}^{(-k)}}^{(k)}, \quad \text{where}$$

$$\hat{d}^{(-k)} = \arg\max_i \left(\frac{1}{K-1} \left(\sum_{\ell=1}^{k-1} x_i^{(\ell)} + \sum_{\ell=k+1}^{K} x_i^{(\ell)} \right) \right) = \arg\max_i \left(\sum_{\ell \neq k} x_i^{(\ell)} \right).$$

Thus, it suffices to compare $x_{\hat{d}}^{(k)}$ and $x_{\hat{d}^{(-k)}}^{(k)}$. By definitions of $\hat{d}$ and $\hat{d}^{(-k)}$:

$$\sum_{\ell} x_{\hat{d}}^{(\ell)} \geq \sum_{\ell} x_{\hat{d}^{(-k)}}^{(\ell)} \quad \text{and} \quad \sum_{\ell \neq k} x_{\hat{d}^{(-k)}}^{(\ell)} \geq \sum_{\ell \neq k} x_{\hat{d}}^{(\ell)}.$$

Consequently,

$$x_{\hat{d}}^{(k)} = \sum_{\ell} x_{\hat{d}}^{(\ell)} - \sum_{\ell \neq k} x_{\hat{d}}^{(\ell)} \geq \sum_{\ell} x_{\hat{d}^{(-k)}}^{(\ell)} - \sum_{\ell \neq k} x_{\hat{d}^{(-k)}}^{(\ell)} = x_{\hat{d}^{(-k)}}^{(k)}. \qquad \square$$

It can be observed that if $\hat{d}^{(-k)} = \hat{d}$ for all k, then the estimates provided are equal.

1–3.	Perform steps 1–3 of Algorithm 1, including the substeps.
4.	**for** $k := 1$ **to** K, (The indexing loop.)
4.1	Pick the subset size using which maximum performance was obtained on the kth execution of step 3.4: $\hat{d}^{(k)} = \arg\max_i \left(x_i^{(k)} \right)$.
4.2	Record the performance for this very subset size on all the other iterations except the kth.
4.3	Compute the average of the $K - 1$ estimates obtained in step 4.2: $\tilde{x}_{\hat{d}(k)}^{(-k)} = \frac{1}{K-1} \left(\sum_{\ell=1}^{k-1} x_{\hat{d}(k)}^{(\ell)} + \sum_{\ell=k+1}^{K} x_{\hat{d}(k)}^{(\ell)} \right)$.
4.4	For comparison purposes, you can also record the performance for the full set on all the other $K - 1$ iterations.
	end;
5.	Average all the K subset sizes obtained during the different executions of step 4.1. This average is the estimate for the optimal subset size.
6.	Average all the K performance estimates obtained in step 4.3. This average is the estimate for the performance of the best subset having the size discovered during step 5.
7–8.	Perform steps 7–8 of Algorithm 1.

Algorithm 3. The cross-indexing B algorithm. Again, the statistical measure computed in steps 4.3, 5 and 6 need not be the average.

On the other hand, in cross-indexing B outlined in Algorithm 3, the estimate number k for the optimal subset size, $\hat{d}^{(k)}$, is determined in step 4.1 using the estimates obtained during the kth iteration of the evaluation loop. This estimate is then used to look up the performances for the same subset size, but for the other iterations (step 4.3).

The cross-indexing algorithms described produce only a single value for both the optimal subset size and the corresponding performance. However, those estimates that undergo averaging in steps 5 and 6 of Algorithms 2 and 3 could also be used to determine some kind of confidence intervals, or to assess the stability of the solution. Unfortunately, such attempts are outside the scope of this paper.

6   Experiments

This section compares the following four methods for determining the optimal subset size and for assessing the performance of the best subset having that size:

0. No outer loop of cross-validation at all,
1. Outer-loop CV (Algorithm 1),
2. Cross-indexing A (Algorithm 2), and
3. Cross-indexing B (Algorithm 3).

For each dataset, the optimal subset size is determined with every method. Also, the increase in accuracy that can be obtained when the optimal subset of that size is chosen, instead of the full set, is estimated. Then, a classifier is trained using the said optimal subset, and that classifier is used to classify the held-out test data. Doing the same with the full feature set and subtracting gives us the ground-truth improvement due to selecting features. This value can then be compared to the improvement predicted by the estimation approach.

The cardinality of the optimal subset as determined using method i is denoted with $\hat{d}_i$, and this value when divided by the total number of candidate features (D) and multiplied by 100% is signified by η_i. The estimated benefit due to choosing $\hat{d}_i$ features, i.e., the difference between the estimated accuracies using the optimal subset and the full set, is signified by $\delta_i^{(e)}$. On the other hand, the same difference when measured utilizing the held-out test set is denoted using $\delta_i^{(t)}$. To determine the bias of the different approaches, we need to determine the difference between these two differences: $\Delta_i = \delta_i^{(e)} - \delta_i^{(t)}$.

The smaller the absolute value of Δ_i, the smaller the bias in determining the benefits due to choosing the optimal subset found using the corresponding estimation method. Thus, from the viewpoint of this paper, the best method is signified by the smallest value of $|\Delta_i|$.

To estimate the standard deviations of the said key figures, every experiment is repeated 30 times with a different seed for the random number generator.

In the context of the 1NN classifier, the type of the inner cross-validation loop is varied: namely, LOOCV and 5-fold CV are used. However, LOOCV gets computationally too expensive with the C4.5 induction algorithm: therefore, such experiments are not done. The outer CV or cross-indexing loop always uses 5 folds.

6.1   Datasets

The datasets used in the experiments are summarized in Table 1. Each of them is publicly available at the UCI Machine Learning Repository.[2]

Table 1. The datasets used in the experiments. The number of features in the set is denoted by D. One fth of the samples are used during the search (see text). The classwise distribution of the samples in the original set is shown in the next column, and the number of training samples used (roughly the total number of samples divided by f) is given in the last column, denoted by m.

dataset	D	f	samples	m
dermatology	33	2	20–112 (total 366)	184
ionosphere	34	2	126 and 225	176
mushroom	112	10	3916 and 4208	813
sonar	60	2	97 and 111	105
spectf	44	2	95 and 254	175
waveform	40	5	1653–1692 (total 5000)	1000
wdbc	29	2	212 and 357	284
wpbc	32	2	47 and 151	99

The mushroom dataset in the repository has 21 categorical features for which no value is missing. In the experiments of this paper, 112 binary features are generated from them using 1-of-N coding: a categorical feature having N possible values will generate N binary features, such that the jth of these is assigned the value 1 if the sample, according to this feature, belongs to the jth category, and 0 otherwise. The mushroom set is chosen because it has previously expressed interesting behavior in the context of feature selection [13].

Before doing anything, each dataset is divided into the set to be used during the search, and the held-out independent test set. It is this division and all the subsequent steps that are — for each combination being tested — repeated for 30 times. The split is controlled using the parameter f (see Table 1): the dataset is first divided into f subsets, of which one is chosen as the set to be used during the search while the other $f - 1$ subsets constitute the hold-out set. Note that this is not related to any of the different levels of cross-validation: the purpose of the parameter f is just to make sure that the training sets do not get prohibitively large in those cases where the dataset happens to have a lot of samples. CV is then done during the search — potentially in two nested loops — in order to be able to guide the selection towards the useful feature subsets, and to estimate their benefits.

6.2   Results

For clarity, the figures introduced in the beginning of this section ($\hat{d}_i$, η_i, $\delta_i^{(e)}$, $\delta_i^{(t)}$ and Δ_i) are first shown in Table 2 for a single dataset using the 1NN classifier and the SFS search algorithm. Then, more results are lined up: Table 3 contains the essential

---

[2] http://www.ics.uci.edu/~mlearn/MLRepository.html

Table 2. Results obtained for the `ionosphere` dataset using the 1NN classifier and the SFS algorithm. The different values of i refer to the different approaches: no outer loop of CV at all ($i = 0$), outer-loop CV as in Algorithm 1 ($i = 1$), cross-indexing A ($i = 2$), and cross-indexing B ($i = 3$). Smaller (absolute) value of $\Delta_i = \delta_i^{(e)} - \delta_i^{(t)}$ implies less observed bias, and thus a better method. The '$\pm$' signifies a single standard deviation.

inner CV	i	$\hat{d}_i$	η_i (%)	$\delta_i^{(e)}$	$\delta_i^{(t)}$	Δ_i
LOO	0	6 ± 4	19 ± 13	14 ± 3	2 ± 3	12 ± 4
	1	9 ± 5	25 ± 15	5 ± 2	2 ± 2	3 ± 3
	2	9 ± 5	25 ± 13	2 ± 3	2 ± 3	-1 ± 4
	3	8 ± 3	23 ± 9	2 ± 3	2 ± 3	-0 ± 4
5-fold	0	7 ± 4	21 ± 11	11 ± 3	2 ± 3	9 ± 3
	1	13 ± 9	39 ± 26	5 ± 3	1 ± 3	4 ± 4
	2	11 ± 6	32 ± 16	2 ± 4	1 ± 3	0 ± 5
	3	9 ± 4	26 ± 12	1 ± 3	2 ± 3	-0 ± 4

Table 3. Results like those in Table 2 but for all the datasets, as obtained using the SFS algorithm and the 1NN classifier architecture. Again, a smaller absolute value of Δ_i suggests that the estimation method is less biased, thus better.

dataset	inner CV	η_0	η_1	η_2	η_3	Δ_0	Δ_1	Δ_2	Δ_3
dermatology	LOO	52 ± 26	68 ± 24	68 ± 21	49 ± 15	4 ± 2	2 ± 2	0 ± 3	-0 ± 3
	5-fold	53 ± 16	69 ± 22	73 ± 16	48 ± 11	4 ± 2	2 ± 2	1 ± 2	0 ± 3
ionosphere	LOO	19 ± 13	25 ± 15	25 ± 13	23 ± 9	12 ± 4	3 ± 3	-1 ± 4	-0 ± 4
	5-fold	21 ± 11	39 ± 26	32 ± 16	26 ± 12	9 ± 3	4 ± 4	0 ± 5	-0 ± 4
mushroom	LOO	26 ± 20	69 ± 26	76 ± 21	30 ± 11	1 ± 1	0 ± 0	-0 ± 0	-1 ± 2
	5-fold	13 ± 8	60 ± 24	57 ± 24	18 ± 8	1 ± 2	0 ± 0	-0 ± 0	-2 ± 4
sonar	LOO	50 ± 18	66 ± 17	65 ± 16	53 ± 10	17 ± 7	5 ± 5	0 ± 6	-2 ± 7
	5-fold	38 ± 16	61 ± 18	61 ± 12	46 ± 10	18 ± 6	4 ± 6	-2 ± 7	-2 ± 7
spectf	LOO	32 ± 14	33 ± 23	37 ± 21	31 ± 12	18 ± 5	7 ± 6	3 ± 8	2 ± 6
	5-fold	30 ± 19	36 ± 27	35 ± 20	32 ± 13	16 ± 3	8 ± 4	3 ± 6	2 ± 5
waveform	LOO	52 ± 13	63 ± 17	63 ± 14	56 ± 9	6 ± 1	2 ± 2	1 ± 2	-0 ± 2
	5-fold	46 ± 11	58 ± 17	60 ± 14	53 ± 11	5 ± 2	2 ± 2	1 ± 2	0 ± 2
wdbc	LOO	35 ± 27	75 ± 28	78 ± 19	60 ± 15	4 ± 3	1 ± 2	0 ± 3	-0 ± 3
	5-fold	44 ± 21	68 ± 26	67 ± 20	49 ± 15	5 ± 2	2 ± 3	0 ± 4	-0 ± 3
wpbc	LOO	37 ± 30	77 ± 29	76 ± 22	56 ± 17	17 ± 11	6 ± 8	2 ± 10	0 ± 9
	5-fold	27 ± 20	66 ± 32	62 ± 25	48 ± 13	19 ± 10	7 ± 9	2 ± 9	1 ± 9

information of Table 2, but for all the datasets. For brevity, only η_i and Δ_i are now shown. Tables 4 and 5 report the results for the combinations C4.5/SFS and 1NN/SFFS, respectively.

6.3 Discussion

Table 2 shows us directly that for the `ionosphere` dataset, the estimates provided by method 0 — no outer loop of cross-validation at all — have a significant amount of optimistic bias, which is not a new result [1]. As expected, the straightforward outer-loop CV (approach 1) clearly lessens the problem, but does not nullify it. On the other hand, it seems that both cross-indexing methods are able to make the bias effectively vanish.

Based on Table 3, it can be readily observed that also for the other datasets, the bias incurred by cross-indexing (Δ_2 or Δ_3) is much smaller than that caused by the other approaches (Δ_0 and Δ_1). This difference really does make a difference when we want to estimate the degree of improvement (in the accuracy of a classifier) that can be attained by running a feature selection algorithm.

The `mushroom` dataset deserves some special attention. While the outer-loop CV and cross-indexing A are basically able to report perfectly unbiased accuracy results for it, those methods fail to identify the fact that virtually error-free results can be obtained with much less than half the number of features. Although the results due to cross-indexing B are slightly biased to the negative direction, it is able to report a much smaller optimal feature subset size, which already separates the classes extremely well.

Table 4. Results calculated using the SFS strategy together with the C4.5 classifier

dataset	inner CV	η_0	η_1	η_2	η_3	Δ_0	Δ_1	Δ_2	Δ_3
dermatology	5-fold	42 ± 19	66 ± 25	61 ± 22	37 ± 13	4 ± 3	2 ± 2	1 ± 2	-0 ± 3
ionosphere	5-fold	34 ± 21	62 ± 33	56 ± 28	34 ± 14	6 ± 3	2 ± 2	-1 ± 3	-1 ± 3
mushroom	5-fold	20 ± 14	40 ± 25	38 ± 23	18 ± 10	1 ± 0	0 ± 0	0 ± 0	-0 ± 0
sonar	5-fold	24 ± 15	29 ± 25	35 ± 20	27 ± 12	17 ± 8	6 ± 6	0 ± 9	0 ± 8
spectf	5-fold	35 ± 21	61 ± 32	57 ± 25	42 ± 14	13 ± 6	4 ± 5	-0 ± 4	-1 ± 5
waveform	5-fold	36 ± 17	52 ± 25	52 ± 19	41 ± 12	5 ± 2	2 ± 2	-0 ± 2	-0 ± 2
wdbc	5-fold	30 ± 17	44 ± 30	44 ± 26	27 ± 13	4 ± 3	2 ± 2	0 ± 2	-0 ± 2
wpbc	5-fold	42 ± 23	21 ± 21	22 ± 19	23 ± 13	9 ± 8	2 ± 7	0 ± 7	1 ± 8

Table 5. Results for the SFFS algorithm and the 1NN classifier

dataset	inner CV	η_0	η_1	η_2	η_3	Δ_0	Δ_1	Δ_2	Δ_3
dermatology	LOO	34 ± 10	77 ± 25	72 ± 20	50 ± 11	6 ± 3	2 ± 2	1 ± 3	1 ± 3
	5-fold	51 ± 20	74 ± 22	72 ± 18	48 ± 10	5 ± 4	2 ± 3	1 ± 3	-0 ± 3
ionosphere	LOO	23 ± 11	26 ± 20	28 ± 18	23 ± 8	13 ± 5	3 ± 4	-1 ± 5	-1 ± 5
	5-fold	21 ± 9	36 ± 25	34 ± 18	28 ± 12	9 ± 4	3 ± 3	-1 ± 4	-0 ± 3
mushroom	LOO	7 ± 2	77 ± 27	67 ± 24	20 ± 9	4 ± 12	0 ± 0	-0 ± 0	-1 ± 1
	5-fold	9 ± 6	60 ± 26	55 ± 23	20 ± 10	2 ± 7	0 ± 0	-0 ± 0	-2 ± 6
sonar	LOO	26 ± 7	62 ± 26	65 ± 19	47 ± 15	21 ± 8	7 ± 7	2 ± 7	-0 ± 7
	5-fold	39 ± 15	63 ± 21	59 ± 15	43 ± 10	20 ± 7	6 ± 6	-0 ± 6	-1 ± 5
spectf	LOO	42 ± 13	43 ± 31	41 ± 25	33 ± 13	20 ± 6	7 ± 7	-0 ± 7	-0 ± 5
	5-fold	34 ± 18	33 ± 26	36 ± 19	33 ± 12	16 ± 4	6 ± 5	1 ± 5	1 ± 5
waveform	LOO	49 ± 9	55 ± 20	53 ± 13	50 ± 10	7 ± 2	2 ± 2	0 ± 2	0 ± 2
	5-fold	49 ± 14	58 ± 17	58 ± 15	52 ± 9	6 ± 2	2 ± 2	0 ± 2	-1 ± 2
wdbc	LOO	18 ± 7	75 ± 26	70 ± 21	47 ± 15	4 ± 3	1 ± 2	0 ± 2	-1 ± 2
	5-fold	54 ± 20	73 ± 23	71 ± 17	46 ± 13	4 ± 3	1 ± 2	-1 ± 3	-1 ± 3
wpbc	LOO	13 ± 6	64 ± 38	64 ± 31	50 ± 18	20 ± 9	7 ± 6	2 ± 7	1 ± 7
	5-fold	37 ± 22	61 ± 31	57 ± 22	43 ± 14	21 ± 9	9 ± 8	2 ± 8	3 ± 8

In general, it can be observed that the estimates given for the size of the optimal subset are surprisingly close to each other for the outer-loop CV (method 1) and cross-indexing A (method 2) — it is just that the cross-indexing approach gives a less biased estimate for the performance. On the other hand, it appears that cross-indexing B (method 3) can identify equally performing subsets that, on the average, tend to be somewhat smaller.

Finally, Tables 4 and 5 reveal that the observations made are not too dependent on a particular choice of classifier architecture or subset selection strategy.

7 Summary

In feature selection, a practitioner typically needs to know the optimal feature subset size for a given dataset, and how much choosing the optimal subset of that size increases the performance.

Traditionally, cross-validation is used to give minimally biased results while using the data available as effectively as possible. However, when cross-validation is done in evaluating and comparing the models, an outer loop of cross-validation is needed for assessing the winner model. Indeed, an outer loop has been suggested for measuring the benefits due to feature selection.

In this paper, it is shown that a simple implementation of an outer cross-validation loop still gives biased estimates for the accuracy of the optimal subset as compared to the full set comprised of all the features. To tackle this problem, a new approach called "cross-indexing" is introduced in the form of two algorithms. They require practically no extra computation, nevertheless are able to give superior, virtually unbiased estimates.

References

[1] J. Reunanen. A pitfall in determining the optimal feature subset size. In *Proc. of the 4th Int. Workshop on Pattern Recognition in Information Systems (PRIS 2004)*, pages 176–185, Porto, Portugal, 2004.

[2] R. J. Schalkoff. *Pattern Recognition: Statistical, Structural and Neural Approaches*. John Wiley & Sons, Inc., 1992.

[3] P. A. Devijver and J. Kittler. *Pattern Recognition: A Statistical Approach*. Prentice–Hall International, 1982.

[4] J. R. Quinlan. *C4.5: Programs for Machine Learning*. Morgan Kaufmann, 1993.

[5] G. H. John, R. Kohavi, and K. Pfleger. Irrelevant features and the subset selection problem. In *Proc. of the 11th Int. Conf. on Machine Learning (ICML-94)*, pages 121–129, New Brunswick, NJ, USA, 1994.

[6] I. Guyon and A. Elisseeff. An introduction to variable and feature selection. *Journal of Machine Learning Research*, 3:1157–1182, 2003.

[7] M. Stone. Cross-validatory choice and assessment of statistical predictions. *Journal of the Royal Statistical Society*, 36(2):111–133, 1974.

[8] R. Kohavi. A study of cross-validation and bootstrap for accuracy estimation and model selection. In *Proc. of the 14th Int. Joint Conf. on Artificial Intelligence (IJCAI-95)*, pages 1137–1143, Montreal, Canada, 1995.

[9] A. W. Whitney. A direct method of nonparametric measurement selection. *IEEE Transactions on Computers*, 20(9):1100–1103, 1971.

[10] P. Pudil, J. Novovičová, and J. Kittler. Floating search methods in feature selection. *Pattern Recognition Letters*, 15(11):1119–1125, 1994.

[11] P. Somol, P. Pudil, J. Novovičová, and P. Paclík. Adaptive floating search methods in feature selection. *Pattern Recognition Letters*, 20(11–13):1157–1163, 1999.

[12] D. D. Jensen and P. R. Cohen. Multiple comparisons in induction algorithms. *Machine Learning*, 38(3):309–338, 2000.

[13] J. Reunanen. Overfitting in making comparisons between variable selection methods. *Journal of Machine Learning Research*, 3:1371–1382, 2003.

Author Index

Lecture Notes in Computer Science

For information about Vols. 1–3893

please contact your bookseller or Springer

Vol. 3944: J. Quiñonero-Candela, I. Dagan, B. Magnini, F. d'Alché-Buc (Eds.), Machine Learning Challenges. XIII, 462 pages. 2006. (Sublibrary LNAI).

Vol. 3943: N. Guelfi, A. Savidis (Eds.), Rapid Integration of Software Engineering Techniques. X, 289 pages. 2006.

Vol. 3942: Z. Pan, R. Aylett, H. Diener, X. Jin, S. Göbel, L. Li (Eds.), Technologies for E-Learning and Digital Entertainment. XXV, 1396 pages. 2006.

Vol. 3940: C. Saunders, M. Grobelnik, S. Gunn, J. Shawe-Taylor (Eds.), Subspace, Latent Structure and Feature Selection. X, 209 pages. 2006.

Vol. 3939: C. Priami, L. Cardelli, S. Emmott (Eds.), Transactions on Computational Systems Biology IV. VII, 141 pages. 2006. (Sublibrary LNBI).

Vol. 3936: M. Lalmas, A. MacFarlane, S. Rüger, A. Tombros, T. Tsikrika, A. Yavlinsky (Eds.), Advances in Information Retrieval. XIX, 584 pages. 2006.

Vol. 3935: D. Won, S. Kim (Eds.), Information Security and Cryptology - ICISC 2005. XIV, 458 pages. 2006.

Vol. 3934: J.A. Clark, R.F. Paige, F.A. C. Polack, P.J. Brooke (Eds.), Security in Pervasive Computing. X, 243 pages. 2006.

Vol. 3933: F. Bonchi, J.-F. Boulicaut (Eds.), Knowledge Discovery in Inductive Databases. VIII, 251 pages. 2006.

Vol. 3931: B. Apolloni, M. Marinaro, G. Nicosia, R. Tagliaferri (Eds.), Neural Nets. XIII, 370 pages. 2006.

Vol. 3930: D.S. Yeung, Z.-Q. Liu, X.-Z. Wang, H. Yan (Eds.), Advances in Machine Learning and Cybernetics. XXI, 1110 pages. 2006. (Sublibrary LNAI).

Vol. 3929: W. MacCaull, M. Winter, I. Düntsch (Eds.), Relational Methods in Computer Science. VIII, 263 pages. 2006.

Vol. 3928: J. Domingo-Ferrer, J. Posegga, D. Schreckling (Eds.), Smart Card Research and Advanced Applications. XI, 359 pages. 2006.

Vol. 3927: J. Hespanha, A. Tiwari (Eds.), Hybrid Systems: Computation and Control. XII, 584 pages. 2006.

Vol. 3925: A. Valmari (Ed.), Model Checking Software. X, 307 pages. 2006.

Vol. 3924: P. Sestoft (Ed.), Programming Languages and Systems. XII, 343 pages. 2006.

Vol. 3923: A. Mycroft, A. Zeller (Eds.), Compiler Construction. XIII, 277 pages. 2006.

Vol. 3922: L. Baresi, R. Heckel (Eds.), Fundamental Approaches to Software Engineering. XIII, 427 pages. 2006.

Vol. 3921: L. Aceto, A. Ingólfsdóttir (Eds.), Foundations of Software Science and Computation Structures. XV, 447 pages. 2006.

Vol. 3920: H. Hermanns, J. Palsberg (Eds.), Tools and Algorithms for the Construction and Analysis of Systems. XIV, 506 pages. 2006.

Vol. 3918: W.K. Ng, M. Kitsuregawa, J. Li, K. Chang (Eds.), Advances in Knowledge Discovery and Data Mining. XXIV, 879 pages. 2006. (Sublibrary LNAI).

Vol. 3917: H. Chen, F.Y. Wang, C.C. Yang, D. Zeng, M. Chau, K. Chang (Eds.), Intelligence and Security Informatics. XII, 186 pages. 2006.

Vol. 3916: J. Li, Q. Yang, A.-H. Tan (Eds.), Data Mining for Biomedical Applications. VIII, 155 pages. 2006. (Sublibrary LNBI).

Vol. 3915: R. Nayak, M.J. Zaki (Eds.), Knowledge Discovery from XML Documents. VIII, 105 pages. 2006.

Vol. 3914: A. Garcia, R. Choren, C. Lucena, P. Giorgini, T. Holvoet, A. Romanovsky (Eds.), Software Engineering for Multi-Agent Systems IV. XIV, 255 pages. 2006.

Vol. 3910: S.A. Brueckner, G.D.M. Serugendo, D. Hales, F. Zambonelli (Eds.), Engineering Self-Organising Systems. XII, 245 pages. 2006. (Sublibrary LNAI).

Vol. 3909: A. Apostolico, C. Guerra, S. Istrail, P. Pevzner, M. Waterman (Eds.), Research in Computational Molecular Biology. XVII, 612 pages. 2006. (Sublibrary LNBI).

Vol. 3908: A. Bui, M. Bui, T. Böhme, H. Unger (Eds.), Innovative Internet Community Systems. VIII, 207 pages. 2006.

Vol. 3907: F. Rothlauf, J. Branke, S. Cagnoni, E. Costa, C. Cotta, R. Drechsler, E. Lutton, P. Machado, J.H. Moore, J. Romero, G.D. Smith, G. Squillero, H. Takagi (Eds.), Applications of Evolutionary Computing. XXIV, 813 pages. 2006.

Vol. 3906: J. Gottlieb, G.R. Raidl (Eds.), Evolutionary Computation in Combinatorial Optimization. XI, 293 pages. 2006.

Vol. 3905: P. Collet, M. Tomassini, M. Ebner, S. Gustafson, A. Ekárt (Eds.), Genetic Programming. XI, 361 pages. 2006.

Vol. 3904: M. Baldoni, U. Endriss, A. Omicini, P. Torroni (Eds.), Declarative Agent Languages and Technologies III. XII, 245 pages. 2006. (Sublibrary LNAI).

Vol. 3903: K. Chen, R. Deng, X. Lai, J. Zhou (Eds.), Information Security Practice and Experience. XIV, 392 pages. 2006.

Vol. 3902: R. Kronland-Martinet, T. Voinier, S. Ystad (Eds.), Computer Music Modeling and Retrieval. XI, 275 pages. 2006.

Vol. 3901: P.M. Hill (Ed.), Logic Based Program Synthesis and Transformation. X, 179 pages. 2006.

Vol. 3900: F. Toni, P. Torroni (Eds.), Computational Logic in Multi-Agent Systems. XVII, 427 pages. 2006. (Sublibrary LNAI).

Vol. 3899: S. Frintrop, VOCUS: A Visual Attention System for Object Detection and Goal-Directed Search. XIV, 216 pages. 2006. (Sublibrary LNAI).

Vol. 3898: K. Tuyls, P.J. 't Hoen, K. Verbeeck, S. Sen (Eds.), Learning and Adaption in Multi-Agent Systems. X, 217 pages. 2006. (Sublibrary LNAI).

Vol. 3897: B. Preneel, S. Tavares (Eds.), Selected Areas in Cryptography. XI, 371 pages. 2006.

Vol. 3896: Y. Ioannidis, M.H. Scholl, J.W. Schmidt, F. Matthes, M. Hatzopoulos, K. Boehm, A. Kemper, T. Grust, C. Boehm (Eds.), Advances in Database Technology - EDBT 2006. XIV, 1208 pages. 2006.

Vol. 3895: O. Goldreich, A.L. Rosenberg, A.L. Selman (Eds.), Theoretical Computer Science. XII, 399 pages. 2006.

Vol. 3894: W. Grass, B. Sick, K. Waldschmidt (Eds.), Architecture of Computing Systems - ARCS 2006. XII, 496 pages. 2006.